T0251029

Cell Intercommunication

Editor

Walmor C. De Mello
Department of Pharmacology
University of Puerto Rico
San Juan, Puerto Rico

CRC Press
Taylor & Francis Group
Boca Raton London New York

CRC Press is an imprint of the
Taylor & Francis Group, an **informa** business

First published 1990 by CRC Press
Taylor & Francis Group
6000 Broken Sound Parkway NW, Suite 300
Boca Raton, FL 33487-2742

Reissued 2018 by CRC Press

© 1990 by CRC Press, Inc.
CRC Press is an imprint of Taylor & Francis Group, an Informa business

No claim to original U.S. Government works

This book contains information obtained from authentic and highly regarded sources. Reasonable efforts have been made to publish reliable data and information, but the author and publisher cannot assume responsibility for the validity of all materials or the consequences of their use. The authors and publishers have attempted to trace the copyright holders of all material reproduced in this publication and apologize to copyright holders if permission to publish in this form has not been obtained. If any copyright material has not been acknowledged please write and let us know so we may rectify in any future reprint.

Library of Congress Cataloging-in-Publication Data

Cell intercommunication / Walmor C. De Mello.
     p. cm.
  Includes bibliographical references.
  ISBN 0-8493-6257-1
  1. Cell interaction. I. De Mello, Walmor, C.
[DNLM: 1. Cell Communication. 2. Intercellular junctions. QH
604.2 C3935]
QH604.2.C44443   1990
574.87'6--dc20
DNLM/DLC
for Library of Congress   89-22195

Publisher's Note
The publisher has gone to great lengths to ensure the quality of this reprint but points out that some imperfections in the original copies may be apparent.

Disclaimer
The publisher has made every effort to trace copyright holders and welcomes correspondence from those they have been unable to contact.

ISBN 13: 978-1-315-89136-1 (hbk)
ISBN 13: 978-1-351-07046-1 (ebk)

Visit the Taylor & Francis Web site at http://www.taylorandfrancis.com and the
CRC Press Web site at http://www.crcpress.com

PREFACE

The development of a complex organism is the result of an evolutionary scheme that starts with the synthesis of macromolecules such as DNA, the appearance of unicellular organisms, and the capacity of isolated cells to recognize each other and comunicate in such a way that tissues and organs are formed.

In this long series of events, intercellular channels represent a simple and reliable process of intercellular communication that has been preserved throughout evolution.

The role of gap junctions in electrical synchronization of excitable tissue, growth, differentiation, and cancer is presented in this volume with the purpose of informing the reader and providing the young investigator with clear perspectives for future work.

The molecular organization of gap junctions is also discussed, and this offers an opportunity for close inspection of molecular mechanisms involved in regulation of junctional permeability.

Some Newtonian models of cell communication are presented, and the influence of toxicological factors on junctional communication is reviewed.

I hope this volume will be of help to those interested in the process of cell communication and its implications in cell biology, physiology, pharmacology, and oncology. I want to thank the colleagues who joined me in this project and the staff of CRC Press for their help in the preparation of this book.

Walmor De Mello
San Juan, Puerto Rico

EDITOR

Dr. Walmor C. De Mello, M.D., Ph.D., did his post-doctoral training at the State University of New York, Downtown Medical Center, and the National Institute for Medical Research (Mill Hill) in London.

He is a member of the American Physiological Society, Biophysical Society, Society of Neurosciences, and the International Society for Heart Research, as well as the Deutsche Physiologische Gesellschaft. He was a Rockefeller Foundation Fellow at Mill Hill and a Roche Foundation Fellow at the University of Bern, Switzerland.

Dr. De Mello has been involved in the study of cell comunication for a number of years.

CONTRIBUTORS

Grant Carrow, Ph.D.
Graduate Department of Biochemistry
Brandeis University
Waltham, Massachusetts

Chia-Cheng Chang
Department of Pediatrics and Human
    Development
Michigan State University
East Lansing, Michigan

Ida Chow, Ph.D.
Assistant Professor
Department of Biology
The American University
Washington, D.C.

David M. Larson, Ph.D.
Assistant Professor
Cardiovascular Pathology Laboratory
Mallory Institute of Pathology
Boston University School of Medicine
Boston, Massachusetts

Irwin B. Levitan, Ph.D.
Professor
Graduate Department of Biochemistry
Brandeis University
Waltham, Massachusetts

Burra V. Madhukaar
Department of Pediatrics and Human
    Development
Michigan State University
East Lansing, Michigan

Michael A. Mancini, B.S.
Graduate Student
Department of Cellular and Structural
    Biology
University of Texas Health Science
    Center
San Antonio, Texas

Saw Yin Oh
Department of Clinical Pharmacology
Flinders University
Bedford Park, South Australia

Michael J. Olson, Ph.D.
Staff Research Scientist
Biomedical Science Department
General Motors Research Laboratories
Warren, Michigan

Arun K. Roy, Ph.D.
Professor and Head
Division of Molecular Genetics
Department of Obstetrics and
    Gynecology
University of Texas Health Science
    Center
San Antonio, Texas

Otto Traub
Institute of Genetics
Division of Molecular Genetics
University of Bonn
Bonn, Federal Republic of Germany

James E. Trosko, Ph.D.
Professor
Department of Pediatrics and Human
    Development
Michigan State University
East Lansing, Michigan

**Manjeri A. Venkatachalam,
    M.B.B.S.**
Professor
Department of Pathology
University of Texas Health Science
    Center
San Antonio, Texas

Frank Welsch, D.V.M.
Department of Experimental
  Pathology and Toxicology
Chemical Industry Institute of
  Toxicology
Research Triangle Park, North
  Carolina

Klaus Willecke
Institute of Genetics
Division of Molecular Genetics
University of Bonn
Bonn, Federal Republic of Germany

TABLE OF CONTENTS

Chapter 1

THE WAY CELLS COMMUNICATE

Walmor C. De Mello

TABLE OF CONTENTS

I. CELL COMMUNICATION — AN EVOLUTIONARY VIEW

"Whenever we look, we find evolution, diversification, and instabilities."

Jlya Prigogine
Nobel Laureate

Corliss,[1] based on extensive submarine research, suggests that life originated in shallow Archean seas in consequence of the eruption of magma through cracks on the earth's crust. According to this idea submarine hot springs provided the conditions for the generation of life about 3.6 billion years ago.

The biological systems are situated above the physicochemical/matter-energy systems and below the sociocultural systems which together form a theoretical pyramid used by Laszlo[2] to describe the realms of evolution.

It has been suggested that biological evolution occurred by symbiosis of prokaryotic cells into one another to form eukaryotic cells. Margoulis envisages that the first eukaryotic cell was generated when a microbe capable only of anaerobic fermentation of glucose to pyruvate established a symbiotic association with an anaerobic prokaryotic cell which must have had the ability to produce cytochromes and to oxidize all foodstuffs into CO_2.[3]

Although no precise information about when this phenomenon occurred is available, it is probable that the Middle-Late Precambrian Boundary marks the appearance of eukaryotic cells.

The formation of colonies of amoebae is probably one of the oldest examples of intercellular communication. It is known, for instance, that isolated cells can keep growing and dividing as long as bacteria present in the extracellular medium are available as food. Under these conditions the self-sufficient cell seems to be an independent unit of life with no apparent need for interacting with other cells.

A substantial decline in the number of bacteria triggers a process characterized by the generation of pulses of cAMP by some cells followed by interaction with cAMP receptors in cells located far away. The wave of cAMP released into the extracellular space leads to activation of cAMP receptors and consequent movement of these cells towards the cAMP pulsing units.[4-6] In this way an aggregation territory is established. This is an example of how fluctuations in the systems elicited by instability in food supply are able to generate organization.[7]

This primitive example of intercellular communication seems to be exclusively related to the exchange of chemical messages through the extracellular medium, since no intercellular junctions have been found in *Dictyostelium discoideum*.[8]

How did the evolution of biological systems generate eukaryotic cells? According to the theory of punctuated equilibria,[9] evolution occurs when the dominant population within a group of species presenting a similar adaptative plan reaches an unstable state. At this juncture, if instability is strong enough, the dominant species is replaced by a peripheral species.

This theory proposes that evolution concerns the survival of an entire species rather than individuals, as originally proposed by Darwin. The destabilization of a previously dominant community of algae (prokaryotic cells) by the appearance of eukaryotic cells created the opportunity for the generation of more prokaryotes and the consequent generation of more specialized eukaryotes. This process is relatively rapid and leads to the phenomenon of *speciation,* or the emergence of new species.[2] The aggregation of eukaryotic cells seems to create the necessity of a more elaborated process of intercellular communication. Here is probably when the intimate contact between cells led to the appearance of intercellular junctions.

Mesozoa, which are more primitive than sponges, Cnidaria, and flatworms, are one of the oldest organisms in which gap junctions can be indentified.[10] In *Ascaris lumbricoides* the primitive giant somatic muscle cells are electrically coupled and healing-over has been described.[11,12]

Of particular interest in the evaluation of evolutionary trends of cell communication is membrane excitability, an essential acquisition which permitted the cell to respond to adequate environmental stimuli with appropriate changes. This basic property was certainly developed before cell communication. Indeed, isolated cells already present this important characteristic. Even in epithelia electrical excitability is observed. In hydromedusae epithelia, for instance, electrical responses that propagate at a velocity of 30 cm/sec can be identified.[13] In siphonores the conduction velocity of electrical pulses in epithelia (50 cm/sec) is similar to that recorded in nerve rings of medusae and the refractory period is identical to that described in vertebrate nerve fibers.[13]

The two basic processes that are essential requirements for intercellular communication — membrane excitability and the ability of cells to generate and receive chemical messages — are preserved throughout the evolution of biological systems.

It is interesting to note that primitive neurons were initially neurosecretory or growth regulatory cells, and only later did their elongated axons become effective cables with the ability to propagate electrical impulses with a high conduction velocity.[14] This stage of the evolutionary process indicates that chemical responsiveness was not totally differentiated. As epithelia are able to respond to specific stimuli with the generation of propagated electrical impulses, what is the reason for the development of nerve cells?

A clear advantage of nerve cells with long axons is that the electrical impulse can be propagated in one specific direction, whereas in epithelia the spread of electrical impulses through gap junctions occurs in all directions simultaneously.

Moreover, it is important to recognize that the further establishment of chemical synapses between the nerve cells made possible the establishment of pathways through which the electrical impulse can be propagated in just one direction.

On analyzing the whole process of evolution it became clear that life is continually searching for novel structures and functions. In the case of intercellular communication, the appearance of specific pathways of communication between cells provided the conditions for the development of complex behavior.

Biological systems are open systems, exchanging energy or matter with their environment; therefore, generation of change is an important characteristic of these systems. The unidirectionality of chemical synapses as well as the synthesis of inhibitory neurotransmitters at nerve endings represented an additional step in the evolutionary process of intercellular pathways. In the human brain, for instance, millions of neurons are activated through specific pathways in a fifth of a second. Here chemical and electrical synapses contribute to brain function, establishing the most complex system of intercellular communication known.

Although the conduction of electrical messages through gap junctions is certainly more economical because it does not require the complex machinery involved in the synthesis and storage of neurotransmitters, the generation of chemical synapses represents an example of how evolution risks the stability of simple components in order to get more highly developed control systems.[2]

It is recognized today that junctional permeability can be modulated by intracellular factors.[15,16] Therefore, it is reasonable to think that preferential pathways of communication can be achieved in systems containing only gap junctions if the concentration of intracellular regulators or the intensity of extracellular influences, as well as the density of receptors, varies in different areas of the tissue.

Although gap junctions were retained in the evolutionary process as a reliable mechanism of cell communication, evolution generates diversity and it is not surprising that significant variations in the molecular organization of the junctions can be detected among different species. Evidence is indeed available that gap junctional proteins are not the same in different tissues,[17,18] emphasizing again the plasticity of the evolutionary process.

During embryologic development gap junctions are widely distributed. Sheridan[19] demonstrated that the injection of dyes into cells of chick embryos is followed by spread of the dye over large areas. Although neural tube cells are communicated, glial cells and neurons are not coupled at the adult stage.

These findings indicate that embryology occasionally uses some simple and reliable processes of cell communication achieved by evolution, but that more sophisticated control systems are added or preferred when the adult stage is reached.

Studies on the proliferation of neural retinal cells of chick embryo, for instance,[20] showed that the number of gap junctions reaches a maximum just before cessation of cell proliferation. This has been interpreted as evidence that a "factor" diffuses from cell to cell, stopping cell proliferation.[20]

In the embryo of the locust, development is characterized by the appearance of intercellular junctions. The gap junctions are expressed transiently between undifferentiated cells, disappearing later on when the cells begin to cluster and establishing the visual units called ommatidia.[21]

In amphibians the gap junctions are probably important during myogenesis because they make possible the spread of electrical current and consequent muscle activation even before innervation is accomplished.[22,23] One must recognize, however, that more studies are required to clarify the role of gap junctions on differentiation. Lawrence and Green[24] showed, for instance, that insect epidermic cells located in adjacent segments involved in different developmental pathways are coupled to one another and that cells inside the segments are also electrically communicated. These findings are not expected if gap junctions are believed to be involved in differentiation.

The development of gap junctions, which seems related to the necessity of cells to share metabolites and electrical information, is completely suppressed in some cells. One reasonable explanation is that cell differentiation provides in some cases a certain degree of metabolic autosufficiency that enables the cells to survive isolated from their neighbors. In skeletal muscle, for instance, the fibers are completely isolated from each other, making possible the establishment of functional units composed of a few fibers and their respective axons. The number of functional units activated varies with the needs of the organism. This particular arrangement is required for the maintenance of posture that requires a constant state of contraction of extensor muscle of the legs, a

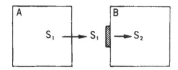

FIGURE 1. Model of cell communication in which cell A releases a chemical signal that produces a change in cell B through receptor (dotted area in B) interaction.

process that is achieved by rotating the number of functional units and consequently avoiding fatigue.

In this particular situation the presence of intercellular junctions between skeletal muscle cells would be incompatible with the organization of isolated functional units as mentioned above. Indeed, the spread of electrical current between the muscle cells would force the whole muscle to respond to a stimulus provided by just one of the axons involved in the innervation of the muscle. This organization contrasts with heart muscle in which intercellular communication is essential for electrical synchronization and development of strong contractions of the ventricular muscle.

Current knowledge of the intricacies of the evolutionary trends of cell communication is limited. Future studies on the comparative biochemistry and physiology of intercellular junctions will certainly provide a better understanding of this important topic.

II. MODELS OF INTERCELLULAR COMMUNICATION

A simple model of cell-to-cell communication is represented by two adjacent cells separated by extracellular fluid. A chemical signal released by cell A can induce changes in cell B if this cell is provided with specific receptors at the level of the surface cell membrane (Figure 1). This is precisely the situation that prevails when colonies of amoebae *(Distyostelium discoideum)* are organized.[6]

It is quite possible that the establishment of chemical synapses between neurons represents a highly elaborated scheme of the simple process of cell communication of amoebae described above. Indeed, the addition of "autopoietic" function (from the Greek for "self-creating"[25]) to the system make possible the continuous renewal of chemical transmitters and synaptic vesicles that are essential for this type of intercellular communication.

Cross-regulation release of two neurotransmitters such as acetylcholine and vasointestinal polypeptide (VIP) has been described in cerebral cortex and rat submandibular gland[26] and represents another example of how evolution utilizes self-regulatory or autocatalytic processes.[7] As discussed above, this type of mechanism of cell communication was necessary for the establishment of pathways of communication between millions of neurons in the central nervous system of vertebrates.

The price for this complex mechanism is, as expected, high. A major dysfunction in any of the stages of the process leads to severe consequences such as Parkinson's disease or myasthenia gravis. On the other hand, the establishment of gap junctions between apposing cells makes possible the quick spread of electrical current in heart

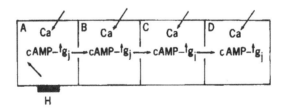

FIGURE 2. In this model of cell communication only cell A presents receptor (dotted area) to a hormone (H). The synthesis of a second messenger (cAMP) in A is followed by increase in junctional conductance (g_j) and diffusion of the nucleotide to cell B in which g_j is also increased. Diffusion of cAMP to cells C and D extends the cascade reaction and makes it possible for cells without receptor to respond to hormonal action. In all cells inward Ca current through surface cell membrane is produced by cAMP.

muscle, dentritic trees, septate axons, epithelia, uterine muscle, and other types of smooth muscle.

Adaptative changes in junctional conductance are probably involved in the maintenance of different values of membrane potential inside the same cell population. In heart, for instance, the membrane potential in pacemaker cells of the sinoatrial node is about 60 mV, whereas in the right atrium values of 75 to 80 mV are found. Although there is a gradation between these values at the border zone, the cells located at the boundary between the two tissues still present a clear difference in their membrane potential. In this particular case, if the junctional conductance prevailing in the pacemaker cells is similar to that found in large Purkinje fibers of the ventricle, for instance, the spread of electrotonic current between pacemaker and atrial cells would increase the membrane potential of pacemaker cells, impairing the effectiveness of these cells to pace the heart.

Evidence exists, however, that the gap junctions between pacemaker cells in the mammalian heart are small in diameter and their number appreciably smaller than that found in atrium or ventricle.[27] These findings might indicate that the maintenance of different values of membrane potential at the boundary between sinoatrial node and right atrium is in part explained by a control of junctional communication in this area.

Certainly other factors, such as the effectiveness of Na/K pump and the different ionic permeability of the nonjunctional cell membrane in pacemaker and atrial cells, contribute to the preservation of the difference between their membrane potentials.

An important model of cell communication is represented by cells coupled through gap junctions, but only a few contain specific receptors in the nonjunctional cell membrane to neurotransmitters (Figure 2). The interaction of the transmitter molecules with the receptor (epinephrine, for instance) increases the intracellular concentration of cAMP through the activation of adenylate cyclase. In some systems the increase of the intracellular levels of cAMP leads to phosphorylation of junctional proteins and a quick increase in junctional conductance (g_j).[16,28] Cyclic AMP not only increases g_j but diffuses easily to nearby cells through gap junctions.[29] If the concentration of cAMP diffusing into the nearby cell devoid of adrenergic receptor is enough to counterbalance the effect of protein kinase inhibitor[16] then the level of phosphorylation of junctional proteins will be increased and g_j enhanced in cell B. The area involved in this cascade

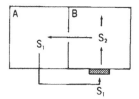

FIGURE 3. Model of intercellular communication showing two cells coupled through a gap junction. Cell A releases a signal (S_1) that activates a receptor in cell B causing the synthesis of an essential messenger (S_2) in this cell. The diffusion of S_2 to cell A might release or inhibit the synthesis of S_1.

reaction will depend on the amount of cAMP that diffuses into cell C or D. According to this model a group of cells without adrenergic receptors on their surfaces will respond to the neurotransmitter indirectly.

The consequences of this model are not limited to the diffusion of cAMP and increase in g_j but extend to the nonjunctional cell membrane of the cells devoid of receptors. In these cells the nucleotide will also increase the current that flows through the voltage-dependent Ca channels in nonjunctional membranes and will influence their excitability or contractility.

This model might be of importance under certain conditions in which the establishment of membrane receptors is suppressed by disease or downregulation. Experimental evidence that cAMP modulates g_j is presented below.

Figure 3 shows a variant of the previous model. Here two cells coupled through gap junctions are interacting in a different way. Cell A releases a signal that activates a receptor located at the surface of cell B and causes the formation of a second messenger inside cell B. This messenger, or some molecules generated by its presence, will diffuse to cell A and will be used appropriately. The molecule S2 might enhance the release of S1 or inhibit its synthesis. If the formation of S1 is increased, the intercellular communication will be greatly increased through an autocatalytic process. If S2 inhibits the synthesis or release of S1 the process will be interrupted and the system will return to its previous steady state. This model will be of value when cell A needs S2 and is not able to synthesize it.

According to this model a biochemical heterogeneity is generated within the system in which some cells play the role of suppliers and others of receivers. In such a system the interruption of junctional communication might represent serious physiological consequences for cell A due to lack of S2, as well as for cell B that is also dependent on S1 for the activation of the receptor and the synthesis of S2.

Electrotonic junctions are frequently observed to occur immediately adjacent to chemical synapses. The meaning of this combination of electrical and chemical junctions is not known.

A possible consequence of this occurrence is presented in a model (Figure 4). The depolarization of the postsynaptic membrane caused by the release of neurotransmitter spreads out back into the presynaptic cell modulating or enhancing the release of the transmitter. A change of less than 1 mV in membrane potential of the presynaptic neuron is enough to modify the synaptic transmission.[30]

This type of feedback between pre- and postsynaptic cells might represent a self-

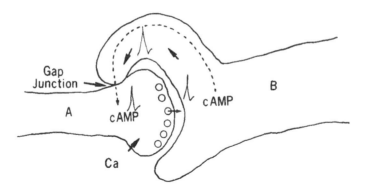

FIGURE 4. Model of cell communication showing modulation of transmitter release (cell A) caused by spread of depolarizing current or cAMP diffusion back into the presynaptic ending through gap junction.

sustaining mechanism of neurotransmitter release — a kind of electric clock.

Moreover, the presence of intercellular channels between the two cells makes possible the flow of chemical information between them. A reasonable possibility is that the generation of cAMP inside the postsynaptic cell diffuses into the presynaptic terminal enhancing the inward Ca current through the surface membrane and consequently increasing the transmitter release.

The coexistence of gap junctions and chemical synapses between neurons might be also involved in a type of negative feedback as presented in Figure 5. According to this model the postsynaptic cell is connected to an intermediary neuron through gap junctions. The establishment of an inhibitory chemical synapse between the presynaptic cell and the intermediary neuron leads to hyperpolarization of the presynaptic neuron every time the postsynaptic cell is excited.

III. JUNCTIONAL PERMEABILITY REGULATION

A central question in gap junction physiology is the meaning of junctional permeability regulation.

The answer to this question is certainly related to the physiological characteristics prevailing in each tissue. The regulation of junctional conductance in cardiac muscle or other excitable tissues, for instance, is supposed to have a different meaning than that of salivary glands or pancreatic cells.

This focuses our attention on a strategic point: the understanding of the physiological meaning of intercellular communication depends on the good use of an intercellular approach and not exclusively on the junction ultrastructure or molecular properties of junctional proteins.

Let us take the cardiac muscle as an example. This is an excitable tissue in which the electrical coupling between the cells is fundamental for the spread of electrical impulses and consequent synchronization of the electrical firing of many cardiac cells. As each cell represents a contractile unit it is easy to visualize that electrical synchronization is required for mechanical synchronization also.

This means that myocardial cells must be always electrically connected in order to

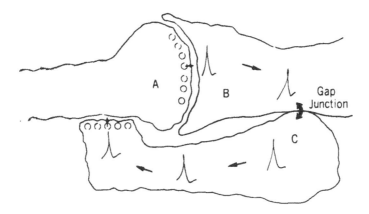

FIGURE 5 In this model of cell communication the postsynaptic cell (B) is connected to an intermediary neuron (C) through gap junctions. The establishment of an inhibitory chemical synapse between the presynaptic cell (A) and the interneuron leads to hyperpolarization of the presynaptic ending any time the postsynaptic cell is excited.

provide the physiological conditions for the heart muscle. In other words, cell decoupling in cardiac muscle is not related to physiological regulation of heart function; on the contrary, cell decoupling is synonymous with heart pathology.

The finding that the intracellular Ca injection leads to cell decoupling in heart[31] as well as in Chironomus salivary gland[32] represents in itself evidence that junctional conductance can be suppressed by an important intracellular factor. Again when the physiological meaning of this finding in heart, for instance, is analyzed, the conclusion is that such an event is not expected to occur under physiological conditions. However, the finding that Ca is a good cell decoupler has pathological significance because during myocardial damage the inward movement of Ca markedly enhances the junctional resistance isolating the normal from the lesioned cells (healing-over) and consequently avoids depolarization of normal tissue.[33,34]

The free $(Ca^{2+})_i$ required to suppress the electrical coupling has been estimated in Chironomus salivary gland and found to be 5 to 8×10^{-5} M.[35]

In cardiac muscle the intracellular free $(Ca^{2+})_i$ needed to supress the electrical coupling was reported to be the same or higher than that required to activate contraction.[36] More recently elegant studies made in isolated cell pairs indicated, however, that 0.2 μM of Ca is enough to abolish the cell-to-cell coupling in heart.[37] As stressed before,[38] the hypothesis that Ca is a physiological modulator of g_j cannot be tested on the basis of Ca concentration necessary to abolish the electrical coupling. It is reasonable to think that cell decoupling in many tissues has no physiological significance at all.

The answer to this question can only be achieved if the minimum Ca concentration needed to change g_j over a physiological range is determined, and that this is indeed a normal occurrence is confirmed.

This task is particularly difficult because (1) the compartmentalization of the cytosol imposes uncertainty on the interpretation of measurements of free $(Ca^{2+})_i$, since no precise information is available on the effective Ca concentration near the gap junctions; and (2) the buffer capacity of the cytosol for Ca avoids variations in Ca concentration that can reach the junctions.

In Prigoginian terms all systems have subsystems which are continually fluctuating.[7] It is then expected that fluctuations in the intracellular concentration of Ca, cAMP, and several other factors might contribute to the modulation of g_j. Changes in concentration of Ca or cAMP inside the cells might elicit variations in g_j of different magnitudes inside the same tissue. It is feasible that changes in number or permeabilities of the junctions or variations in cell size, ability to buffer Ca, or even concentrations of cAMP-dependent protein kinase inhibitor in different areas of the same tissue might lead to regional changes in the degree to which g_j is altered by Ca or cAMP. The suppression of the effect of isoproterenol on g_j caused by dialysis of heart cells with cAMP-dependent protein kinase inhibitor supports this view.[16] This hypothesis certainly provides the possibility of generating preferential pathways of cell communication inside the tissue. On the other hand, it is conceivable that variation in pathways might alter the tissue function.

Although Ca is a cell decoupler in heart, Chironomus salivary glands, and nerve, in some systems the sensitivity to Ca seems to be small. In rat lacrimal cells, for instance, calcium ionophore A23187 increases free $(Ca^{2+})_i$ to levels greater than 1 μM and causes only a transient uncoupling.[39] Moreover, no consistent cell decoupling was seen in this tissue when the free $(Ca^{2+})_i$ was increased.

In embryonic cells of Fundulus the sensitivity to Ca is much less than that of Chironomus salivary gland.[40] Internal perfusion of crayfish septate axons with solutions high in Ca or H ion concentrations had no effect on g_j.[41] Recently, Ramon et al.[42] reported that calmodulin introduced into septate crayfish axons appreciably increases the junctional resistance. This finding might be interpreted as evidence that Ca is an important modulator of g_j in this tissue (for review on calmodulin see References 43 and 44).

One reason for the discrepant views concerning the role of Ca as a modulator of g_j is certainly related to the search for universal cell decouplers instead of junctional modulators.

In 1978 Turin and Warner[45] showed that in embryonic cells exposed to 100% CO_2 the electrical coupling was quickly abolished in a reversible fashion. This suppression of cell coupling was clearly associated with a fall in the intracellular pH. Studies made in fish and amphibian blastomeres showed that the relationship pH_i/g_j is steep and occurs over a pH_i range that may be considered physiological.[46]

The sensitivity of g_j to pH varies in different tissues. In pairs of isolated hepatocytes in which the normal pH_i is 7.4 a decrease in g_j was seen only after pH_i was reduced by 1 unit; around pH 6.3 the curve pH_i/g_j is very steep.[47] Previous studies indeed indicated that g_j in isolated hepatocytes is quite insensitive to CO_2 exposure.[48]

In mammalian cardiac cell pairs, however, the pH_i/g_j is less steep (pK < 7.0).[40] In canine Purkinje fibers the healing-over process that is intimately related to cell decoupling[33] is induced by acidity, but only if the pH is 5.5 or lower.[49]

Recently studies made on isolated cell pairs of guinea pig ventricle indicated that g_j is much less sensitive to protons than to Ca ions[37] which are in agreement with previous experiments on the healing-over process.[49]

IV. IS G_J VOLTAGE DEPENDENT?

The first evidence of asymmetric voltage dependence in gap junctions was described by Furshpan and Potter[50] in the crayfish.

In early embryonic cells of amphibia the injection of current pulses into one cell of a pair increases the input resistance of the injected cell and reduces the transfer resistance measured in the other cell.[51,52] Channel closure is probably related to an effect of transjunctional voltage on a dipole moment.[46]

Although in amphibia embryonic cell g_j is not altered by variations in potential between the cytoplasm and bath,[46] in Chironomus salivary glands g_j is strongly influenced by the membrane potential.[53] In these experiments hyperpolarization of both cells of pairs increases g_j and depolarization decreases it.

In adult rat ventricular cell pairs g_j is voltage and time independent.[16,37,54-56] These findings are indeed expected because in atrial or ventricular muscle impulse propagation occurs equally in both direction. Even under conditions of reduced coupling the ohmic behavior of the junctional membrane seems preserved.[57]

Of particular interest is the recent finding that when the phosphorylation of one hemichannel is reduced by intracellular dialysis of cAMP-dependent protein kinase inhibitor in heart cell pairs rectification of junctional membrane is detected (De Mello, in press). In other systems like the rat lacrimal glands rectification of the junctional membrane has also been reported.[39] Fully open junctions are always voltage insensitive but in almost completely closed junctions voltage dependence was found. In these situations the number of open channels depended on the transjunctional voltage but not on membrane potential.[58]

Evidence is available that the nonjunctional membrane potential does not influence the junctional permeability in mammalian heart trabeculae exposed to 40 or 60 mM potassium solution. In these muscles the cell-to-cell diffusion of Lucifer Yellow CH was the same as in the control ($K_0 - 5$ mM).[59]

V. CYCLIC AMP AND JUNCTIONAL CONDUCTANCE

Ca and cAMP are intimately related in the regulation of cell functions induced by hormones.[60] It is then reasonable to think that the nucleotide might have a role in the control of g_j.

Initial studies indicated that in salivary glands of Drosophila cAMP increases the electrical coupling.[61] In heart cells norepinephrine enhances V_2/V_1.[62]

In mammalian liver cells in culture dB-cAMP induces the transfer of molecules from cell to cell. This phenomenon, which requires hours to become evident, seems due to an increase in number of gap junctional channels.[63]

In heart muscle, theophylline (0.4 mM) — a phosphodiesterase inhibitor — increases the electrical coupling.[64]

When cAMP is injected electrophoretically into intact cardiac cells the electrical coupling is increased within 30 s.[65] This effect of the nucleotide is not related to an increase in resistance of nonjunctional membrane because the input resistance and the time constant of cell membrane measured in the injected cell are not increased; on the contrary, they are slightly reduced.[65]

Epinephrine that increases the intracellular concentration of cAMP in heart also increases the electrical coupling.[66] The effect of this drug on heart muscle represents an important example of how intracellular mechanisms involved in hormonal action are compartmentalized. Epinephrine, for instance, increases the inward Ca current through voltage-dependent channels located at surface cell membrane, enhances the Ca uptake by the sarcoplasmic reticulum, and simultaneously increases the junctional conductance by incrementing the intracellular level of cAMP.[67] This variety of hormonal action is possible by simultaneous activation of a different type of protein kinases (see below).

Additional support to the view that cAMP is an important regulator of g_j came from studies of cell-to-cell diffusion of Lucifer Yellow CH in dog heart muscle. When Lucifer Yellow CH, a fluorescent dye,[68] is introduced into the heart cell with the cut-end method of Imanaga,[69] the longitudinal diffusion of the dye along the heart muscle can be followed by measuring the fluorescence at different distances from the loading area. The experimental points are fitted by a theoretical diffusion curve and the diffusion coefficient (D) is estimated with an interactive computer program for nonlinear regression.

The value of D estimated from Crank's equation for diffusion in a cylinder, assuming no loss of tracer through the surface cell membrane, was found to be $4 \pm 0.63 \times 10^{-7}$ cm²/s for control. When dB-cAMP (5×10^{-4} M) or isoproterenol (10^{-5} M) was added to bath solution the diffusivity of the dye was increased to $2 \pm 0.52 \times 10^{-6}$ cm²/s and $2.4 \pm 0.66 \times 10^{-6}$ cm²/s, respectively.[70] The junctional permeability (P_{nexus}) can be estimated from the following equation:

$$D^{-1} = (P_{nexus} \, hq)^{-1} + D_s^{-1}$$

(from Hodgkin, appendix in Weidmann[72]) where Ds is the diffusivity of the dye in the sarcoplasm, h the distance between gap junctions (82 μm), q the ratio of nexal membrane area to the cross-sectional area of cardiac cell (0.266), and D the diffusion coefficient.

Values of P_{nexus} (which has a dimension of velocity) were 3×10^{-4} cm/s for control and 9.1×10 cm/s for dB-cAMP. In these estimations Ds was taken as 1×10^{-6} cm²/s[70] for dB-cAMP. This assumption is based on the finding that the binding of the dye to cytoplasmic proteins (5.5%) does not change with dB-cAMP or isoproterenol.[70]

The finding that the diffusion coefficient of the dye in the sarcoplasm (Ds) was greater than the average value of D for the whole fiber suggests that the longitudinal diffusion of Lucifer Yellow CH is restricted at gap junctions. Studies on the longitudinal diffusion of ^{42}K,[72] TEA,[73] Procion yellow, 6-carboxyfluorescein, lissamine rhodamine B-200,[69,74] cAMP,[29] and Lucifer Yellow CH[70] are in reasonable agreement with the kinetics of their flow-through restrictive channels.[75] It is important to add that not only the molecular weight but also the charge of the molecule is important in determining their transit through gap junctions.

Negatively charged molecules are less permeable than neutral tracers of same size.[76,77] These findings suggest that the intercellular pore has fixed negative charges. Assuming that this is the case and following the postulate of Einsenman,[78] it can be postulated that during penetration through the pore cations interact with fixed or mobile membrane sites to a degree depending on the anionic field strength of the sites and the

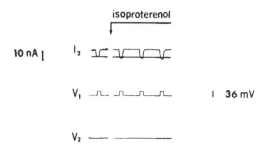

FIGURE 6. Influence of isoproterenol (10^{-6} M) on g_j of a pair of rat ventricular cells. Increase in junctional current (I_2) was seen within 30 s of adding the drug to the bath. Holding potential for both cells was 40 mV. Cell 1 was depolarized by 36 mV (V_1) while the membrane potential of cell 2 was kept unchanged (V_2). Pulse duration was 100 ms. (From De Mello, W. C., *Biochem. Biophys. Res. Commun.*, 154, 509, 1988. With permission.)

size of the cation. The anionic field strength of the pore sites is probably modulated by pH changes (a recent discussion on pH and g_j was presented by Bennett[79]).

The mechanism by which cAMP enhances g_j is not known. It was proposed that the nucleotide activates a cAMP-dependent protein kinase which promotes the phosphorylation of gap-junctional proteins with consequent increase in g_j within seconds.[28,65]

Other studies[63] reported increase or induction of intercellular communication in cell lines exposed to dB-cAMP. This phenomenon, which takes hours to occur, seems related to an increased synthesis of intercellular channels.

Recently, enzymatic techniques for isolating cardiac cells in which the electrical properties of gap junctions can be studied directly were used to investigate the phosphorylation hypothesis.[16]

On these experiments made on isolated cell pairs from rat ventricle the junctional resistance was determined using two separated voltage-clamp circuits. Giga-ohms sealing was achieved in each cell and then the cell membrane was broken and a whole cell clamp was produced. Because each cell was connected to a separate voltage clamp amplifier it was possible to control the nonjunctional membrane potential in each cell as well a the voltage gradient across the intercellular junction. Measurements of g_j were made under control conditions and then isoproterenol (10^{-6} M) — a beta-adrenergic agonist that increases cAMP inside the cell — was added to extracellular fluid. Figures 6 and 7 show the effect of isoproterenol on g_j. As can be seen, the drug increased g_j within 30 s, and a maximal effect of 40% (SE ± 11%; n=10) was found after approximately 60 s.[16]

Previous studies of Saez et al.[80] showed that glucagon increases g_j in pairs of hepatocytes within 2 to 3 min, a finding that supports the cAMP hypothesis.

The effect of isoproterenol on g_j is totally suppressed by dialysis of cAMP-dependent protein kinase inhibitor into both cells of heart cells (see Figures 7 and 8). These observations sustain the hypothesis[28] that the activation of a protein kinase with consequent phosphorylation of gap-junctional proteins is a necessary step in the series of events that led to the increase of g_j caused by cAMP.[16]

The influence of cAMP on g_j varies in different systems. In the horizontal cells of

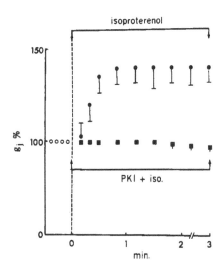

FIGURE 7. Effect of isoproterenol (10^{-6} M) (black circles) on g_j of ten cell pairs under control conditions. Vertical dashed line indicates moment the drug was added to the bath. Control value of g_j taken as 100%. Bottom line (black squares) represents results from 10 cell pairs in which cAMP-dependent protein kinase inhibitor (20 µg/ml) was dialyzed into both cells inhibiting the effect of isoproterenol on g_j. The vertical line at each point is the SE of the mean. (From De Mello, W. C., *Biochem. Biophys. Res. Commun.*, 154, 509, 1988. With permission.)

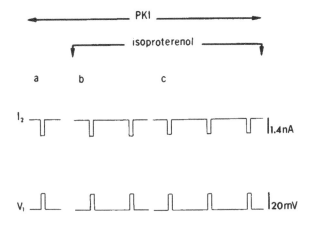

FIGURE 8. Lack of action of isoproterenol (10^{-6} M) on g_j when the protein kinase inhibitor (20 µg/ml) was previously dialyzed into both cells of a pair. (a) Control; (b) $1^1/_2$ min after administration of isoproterenol; (c) 3 min after isoproterenol was added to bath. Pulse duration was 100 ms. (From De Mello, W. C., *Biochem. Biophys. Res. Commun.*, 154, 509, 1988. With permission.)

retina, for instance, dopamine reduces the spread of Lucifer Yellow, suggesting that the neurotransmitter closes the gap junctions between these cells.[81] Moreover, drugs such as amphetamines that stimulate the release of dopamine by nerve endings also reduce the dye coupling.

In uterus muscle isoproterenol, which increases the intracellular concentration of cAMP, seems to decrease the junctional permeability.[82]

It is important to recognize that the nucleotide plays a different role in different cells. Therefore, it is reasonable to expect that the influence of cAMP on g_j in each type of tissue varies according to its physiological role.

Of particular interest is the interaction between Ca and cAMP on g_j. This problem was investigated in intact cardiac Purkinje cells using a double-barreled microelectrode. One barrel was filled with cAMP ($2\ M$ cAMP-Na salt) and the other with $CaCl_2$ ($0.1\ M$) or [ethylene-bis(oxyethylenenitrilo)] tetraacetic acid (EGTA) ($10\ mM$). The results showed that in fibers exposed to high Ca solutions ($6\ mM$) the iontophoretic injection of cAMP caused an initial increase in V_2/V_1 followed by a drastic decrease of the electrical coupling. If EGTA is injected into the same cell at this moment the coupling coefficient was greatly recovered.[83] The decrease in V_2/V_1 caused by cAMP injection when $(Ca^{2+})_0$ was $6\ mM$ seems to indicate that the nucleotide can impair cell-to-cell coupling if too much Ca is moved into the cell. It is known, indeed, that cAMP enhances the inward movement of Ca through the surface cell membrane in several tissues.[84,85]

As the nucleotide increases the Ca uptake by the sarcoplasmic reticulum,[86] the rise in free $(Ca^{2+})_i$ caused by cAMP injection must overcome the buffer capacity of the cytosol for Ca in order to decrease g_j.

An alternative explanation for the decline in V_2/V_1 is that Ca activates a phosphodiesterase[87] and reduces the intracellular concentration of cAMP. It is known, for instance, that calmodulin, an endogenous Ca-binding protein, can modulate the intracellular concentration of cAMP by stimulating the phosphodiesterase activity.[88]

VI. CELL COMMUNICATION AND HEALING-OVER IN SMOOTH MUSCLE

Gap junctions are certainly involved in the spread of electrical activity in some types of smooth muscle such as the circular muscle of the small intestine of guinea pig,[89,90] in which well-formed gap junctions can be identified.

In other types of smooth muscle, such as the intestinal longitudinal muscle of the same animal, gap junctions seem to be extremely small and represented by a few junctional particles.[90]

Despite the lack of organized gap junctions in guinea-pig *Taenia coli*, cable-like properties have been described for this tissue. Indeed, studies of the spread of electrotonic current in *Taenia coli* showed a space constant of 1.5 mm, which is much larger than the length of a single cell.[91]

The question of whether or not gap junctions are essential for electrical communication in this and other types of smooth muscles was recently reviewed by Daniel.[92]

When cardiac muscle is damaged the spread of injury currents is soon interrupted by the formation of a high-resistance barrier at the area of cell damage, preventing the depolarization of normal cells located near the site of injury.[33,93-95]

The healing-over process has been interpreted as being due to the diffusion of Ca through the damaged area with a consequent increase in resistance of the gap junctions located near the lesion.[95]

As healing over and cell decoupling are intimate processes, it was important to discover if the longitudinal muscle of guinea-pig *Taenia coli* is able to heal over despite the absence of organized gap junction. For this purpose the preparation was dissected and immersed in normal oxygenated Krebs solution. The input resistance (V_0/I_0) was measured before and immediately after damage. It was found that the input resistance near the lesion (100 μm) is appreciably increased within 2 to 3 min of damage. The input resistance measured at different distances from the cut end showed values increasing toward the site of lesion.[96] This distribution of input resistance, when compared with theoretical values estimated from cable equations for a sealed and nonsealed cable, indicates the muscle behaves indeed as a sealed cable.

These findings[97] indicate that, despite the lack of organized gap junctions, the closure of single channels represented by the isolated particles described by Gabella[90] is enough to establish a high electrical resistance barrier (healing over), protecting the normal cells from the flow of injury currents.

In uterus muscle gap junctions are not found except at the end of pregnancy where the cells communicate electrically and chemically.[98-100] In uterus at term not only does the junctional impedance fall but the space constant increases from 2.6 to 3.5 mm.[101] Furthermore, high pharmacological doses of estrogen over several days lead to gap junction formation in nonpregnant uterus.[102,103] Tamoxifen, a competitive antagonist of estrogen, prevents gap junction formation by estrogens.[104]

The induction of gap junction formation in uterine muscle by estrogen requires RNA and protein synthesis,[105] suggesting that the synthesis of the intercellular channel subunit is under the control of the hormone.

ACKNOWLEDGMENT

I thank Mrs. Lagnny Jacobo for her word processing expertise. This research was supported by Grant No. HL-34148 from the National Institutes of Health and in part by No. MO786E87 from MBRS.

REFERENCES

1. **Corliss, J. B.,** On the creation of living cells in submarine hot spring flow reactors, Proc. Fifth ISSOL Meet., Eighth Int. Conf. Origin of Life, Berkeley, California, 1986.
2. **Laszlo, E.,** *Evolution: The Gran Synthesis,* Shambhala, New Science Library, Boston, 1987.
3. **Brooks, J. and Shaw, G.,** *Origin and Development of Living Systems,* Academic Press, London, 1973.
4. **Malchow, D. and Gerisch, G.,** Short-time binding and hydrolysis of cyclic 3'-5'-adenosine monophosphate by aggregating dictyostelium cells, Proc. Natl. Acad. Sci.U.S.A., 71, 2423, 1974.
5. **Green, A. A. and Newell, P. C.,** Evidence for the existence of two types of cAMP binding sites in aggregating cells of dictyostelium discoideum, Cell, 6, 129, 1975.
6. **Gerisch, G. and Hess, B.,** Cyclic-AMP controlled oscillations in suspended dictyostelium cells: their relation to morphogenetic cell interactions, Proc. Natl. Acad. Sci. U.S.A., 71, 2118, 1974.

7. **Prigogine, J.,** *Order Out of Chaos,* Shambhala, New Science Library, Boston, 1984.

8. **Johnson, G., Johnson, R., Miller, M., Borysenko, J., and Revel, J. P.,** Do cellular slime molds form intercellular junctions?, *Science,* 197, 1300, 1977.

9. **Eldredge, N. and Gould, S. J.,** Punctuated equilibria: an alternative to Phylogenetic Gradualism, in *Models in Paleobiology,* Schopf, E., Ed., Freeman, Cooper, San Francisco, 1972.

10. **Revel, J. P.,** *The Oldest Multicellular Animal and Its Junctions in Gap Junctions,* Herzberg, E. and Johnson, R., Eds., Alan R. Liss, New York, 1988.

11. **De Mello, W. C.,** The sealing process in heart and other muscle fibers, in *Research in Physiology.* Kao, F. F., Koizumi, D., and Vassalle, M., Eds., Aulo Gaggi, Bologna, 1971, 275.

12. **De Mello, W. C. and Maldonado H.,** Synaptic inhibition and cell communication; impairment of cell-to-cell coupling produced by γ-aminobutyric acid (GABA) in the somatic musculature of *Ascaris lumbricoides, Cell Biol. Intern. Rep.,* 9, 803, 1985.

13. **Mackie, G. O. and Passano, L. M.,** Epithelial conduction in *Hydromedurasae, J. Gen. Physiol.,* 52, 600, 1968.

14. **Bullock, T. H. and Horridge, G. A.,** *Structure and Function in Nervous Systems of Invertebrates,* W. H. Freeman, San Fransicso, 1965.

15. **De Mello, W. C.,** Intercellular communication in cardiac muscle, *Circ. Res.,* 50, 2, 1982.

16. **De Mello, W. C.,** Increase in junctional conductance caused by isoproterenol in heart cell pairs is suppressed by cAMP-dependent protein-kinase inhibitor, *Biochem. Biophys. Res. Commun.,* 164, 509, 1988.

17. **Willecke, K., Traub, O., Look, J., Stutenkemper, R., and Dermietzel, R.,** Different protein components contribute to the structure and function of hepatic gap junctions, in *Gap Junctions,* Hertzberg, E. and Johnson, R., Eds., Alan R. Liss, New York, 1988.

18. **Nilcholson, B. J. and Zhang, J.,** Multiple protein components in a single gap junction: cloning of a second hepatic gap junction protein (MR 21,000), in *Gap Junctions,* Hertzberg, E. and Johnson, R., Eds., Alan R. Liss, New York, 1988.

19. **Sheridan, J. D.,** Electrophysiological evidence for low resistance junctions in early chick embryos, *J. Cell Biol.,* 37, 650, 1968.

20. **Fujisawa, H., Morioka, H., Watanabe, H., and Nakamura, H.,** A decay of gap junction in association with cell differentiation of neural retina in chick embryonic development, *J. Cell Sci.,* 22, 585, 1976.

21. **Eley, S. and Shelton, P. M. J.,** Cell junctions in the developing compound eye of the desert locust *Shistocerca gregoria, J. Embryol. Exp. Morphol.,* 36, 2409, 1976.

22. **Blackshaw, S. E. and Warner, A. E.,** Low resistance junctions between mesoderm cells during development of the trunk muscles, *J. Physiol.,* 255, 231, 1976.

23. **Guthrie, S.,** Intercellular communication in embryos, in *Cell-to-Cell Communication.* De Mello, W. E., Ed., Plenum Press, New York, 1987.

24. **Lawrence, P. A. and Green, S. M.,** The anatomy of a compartment border: the intersequential boundary in Oncopeltus, *J. Cell Biol.,* 65, 373, 1975.

25. **Varela, F., Maturana, H. R., and Uribe, R.,** Autopoiesis: the organization of living systems its characterization and a model, *Bio-Systems,* 5, 65, 1974.

26. **O'Donohue, T. L., Millington, W., Handelmann, G. E., Contreras, P. C., and Cheonwall, B. M.,** On the 50th anniversary of Dale's Law: multiple neurotransmitters neurons, *Trends Pharmacol. Sci.,* 6, 305, 1985.

27. **Masson-Pevet, M.,** *The Fine Structure of Cardiac Pacemaker Cells in the Sinns Node and in Tissue Culture,* Thesis Rodopi Press, Amsterdam, 1979.

28. **De Mello, W. C.,** The role of cAMP and Ca on the modulation of junctional conductance: an integrated hypothesis, *Cell Biol. Intern. Rep.,* 7, 1033, 1983.

29. **Tsien, R. and Weingart, R.,** Inotropic effect of cAMP in calf ventricular muscle studied with a cut-end method, *J. Physiol.,* 260, 117, 1976.

30. **Schmitt, F. O., Paruati, D., and Smith, B. H.,** Electrotonic processing of information by brain cells, *Science,* 193, 114, 1976.

31. **De Mello, W. C.,** Effect of intracellular injection of calcium and strontium on cell communication in heart, *J. Physiol.,* 250, 231, 1975.

32. **Rose, B. and Loewenstein, W. R.,** Calcium ion distribution in cytoplasm visualized by aequorin: diffusion in cytosol restricted by energized sequestering, *Science,* 190, 1204, 1975.

33. De Mello, W. C., The healing-over process in cardiac and other muscle fibers, in *Electrical Phenomena in the Heart*, De Mello, W. C., Ed., Academic Press, New York, 1972, 323.

34. De Mello, W. C., Cell-to-cell communication in heart and other tissues, *Prog. Biophys. Molec. Biol.*, 39, 147, 1982.

35. Oliveira-Castro, G. M. and Loewenstein, W. R., Junctional membrane permeability: effects of divalent cations, *J. Membr. Biol.*, 5, 51 1971.

36. Weingart, R., The action of ouabain on intercellular coupling and conduction velocity in mammalian ventricular muscle, *J. Physiol.*, 264, 341, 1977.

37. Noma, A. and Tsuboi, N., Dependence of junctional conductance on proton, calcium and magnesium ions in cardiac paired cells of guinea pig, *J. Physiol.*, 382, 193, 1987.

38. De Mello, W. C., Modulation of junctional permeability, in *Cell-to-Cell Communication*, De Mello, W. C., Ed., Plenum Press, New York, 1987, 29.

39. Trautmann, A., Neyton, J., Randriamanpita, C., and Giaume, C., Role of divalent cations in the control of gap junction conductance in a rat exocrine gland, in *Gap Junctions*, Herzberg, E. and Johnson, R., Eds., Alan R. Liss, New York, 1988.

40. Spray, D. and Bennett, M. V. L., Physiology and pharmacology of gap junctions, *Annu. Rev. Physiol.*, 47, 281, 1985.

41. Johnson, M. F. and Ramon, F., Voltage independence of an electrotonic synapse, *Biophys. J.*, 39, 115, 1982.

42. Ramon, F., Arellano, O. R., Rivera, H., and Zampighi, G., Effects of internally perfused calmodulin on the junctional resistance of lateral axons, in *Gap Junctions*, Hertzberg, E. and Johnson, R. G., Eds., Alan R. Liss, New York, 1988.

43. Peracchia, C., Permeability and regulation of gap junction channels in cells and in artificial lipid bilayers, in *Cell-to-Cell Communication*, De Mello, W. C., Ed., Plenum Press, New York, 1987.

44. Peracchia, C., The calmodulin hypothesis for the gap junction regulation six years later, in *Gap Junctions*, Hertzberg, E. and Johnson, R., Eds., Alan R. Liss, New York, 1988.

45. Turin, L. and Warner, A. E., Carbon dioxide reversibly abolishes ionic communication between cells of early amphybian embryo, *Nature*, 270, 56, 1978.

46. Spray, D. C., Harris, A. L., and Bennett, M. V. L., Equilibrium properties of a voltage dependent junctional conductance, *J. Gen. Physiol.*, 77, 77, 1981.

47. Spray, D. C., White, R. L., Campos de Carvalho, A. C., Harris, A. L., and Bennett, M. V. L., Gating of gap junction channels, *Biophys. J.*, 45, 219, 1984.

48. Meyer, D. J. and Revel, J. R., CO_2 does not uncouple hepatocytes in rat liver, *Biophys. J.*, 33, 105, 1981.

49. De Mello, W. C., The influence of pH the healing-over of mammalian cardiac muscle, *J. Physiol.*, 339, 299, 1983.

50. Furshpan, E. J. and Potter, D. D., Transmission at the giant motor synapses of the crayfish, *J. Physiol.*, 145, 289, 1959.

51. Spray, D. C., Harris, A. L., and Bennett, M. V. L., Voltage dependence of junctional conductance in early amphibian embryos, *Science*, 204, 432, 1979.

52. Spray, D. C., Harris, A. L., and Bennett, M. V. L., Gap junctional conductance is a simple and sensitive function of pH, *Science*, 211, 712, 1981.

53. Obaid, A. M., Socolar, S. J., and Rose, B., Cell-to-cell channels with two independent regulated gates in series: analysis of junctional channel modulation by membrane potential, calcium and pH, *J. Membr. Biol.*, 73, 69, 1983.

54. Kameyama, M., Electrical coupling between ventricular paired cells isolated from guinea-pig heart, *J. Physiol.*, 336, 345, 1983.

55. White, R. L., Carvalho, A. C., Spray, D. C., Whittenberg, B. A., and Bennett, M. V. L., Gap junctional conductance between isolated pairs of ventricular myocytes from rat, *Biophys. J.*, 41, 217, 1983.

56. Weingart, R., Electrical properties of the nexal membrane studied in rat ventricular cell pairs, *J. Physiol.*, 370, 267, 1986.

57. Maurer, P. and Weingart, R., Cell pairs isolated from adult guinea-pig and rat hearts: effects of $[Ca^{2+}]_i$ on nexal membrane resistance, *Pflüg. Arch.*, 409, 394, 1987.

58. Neyton, J. and Trautmann, A., Physiological modulation of gap junctional permeability, *J. Exp. Biol.*, 124, 93, 1986.

59. **De Mello, W. C. and van Loon, P.,** Junctional permeability in heart muscle is independent upon the non-junctional membrane potential, *Cell Biol. Intern. Rep.,* 11, 1, 1987.

60. **Rasmussen, H.,** Ions as "second messengers", in *Cell Membranes, Biochemistry, Cell Biology and Pathology,* Wesmann, G. and Clairborne, R., Eds., H. P. Publishing, New York, 1975.

61. **Hax, W. M. A., van Venrooij, G. E. P. M., and Vossenberg, J. B. J.,** Cell communication: a cyclic-AMP mediated phenomenon, *J. Membr. Biol.,* 19, 253, 1974.

62. **De Mello, W. C.,** Factors involved in the control of junctional conductance in heart, *Proc. Int. Union Physiol. Sci.,* 12, 319, 1977.

63. **Flagg-Newton, J. L., Dahl, G., and Loewenstein, W. R.,** Cell junctions and cyclic AMP: upregulation of junctional membrane permeability and junctional membrane particles by administration of cyclic nucleotide or phosphodiesterase inhibitor, *J. Membr. Biol.,* 63, 105, 1981.

64. **Estapé, E. and De Mello, W. C.,** Effect of theophylline on the spread of electrotonic activity in heart, *Fed. Proc.,* 41, 1505, 1982.

65. **De Mello, W. C.,** Effect of intracellular injection of cAMP on the electrical coupling of mammalian cardiac cells, *Biochem. Biophys. Res. Commun.,* 119, 1001, 1984.

66. **De Mello, W. C.,** Increased spread of electrotonic potentials during diastolic depolarization in cardiac muscle, *J. Mol. Cell Cardiol.,* 18, 23, 1986.

67. **De Mello, W. C.,** Modulation of junctional permeability, *Fed. Proc.,* 43, 2692, 1984.

68. **Stewart, W. C.,** Functional connections between cells as revealed by dye-coupling with a high fluorescent naphthalimide tracer, *Cell,* 14, 741, 1978.

69. **Imanaga, I.,** Cell-to-cell diffusion of Procion Yellow in sheep and calf Purkinje fibers, *J. Membr. Biol.,* 16, 381, 1974.

70. **De Mello, W. C. and van Loon, P.,** Influence of cyclic nucleotides on junctional permeability in atrial muscle, *J. Mol. Cell Cardiol.,* 19, 83, 1987.

71. **De Mello, W. C. and van Loon, P.,** Further studies on the influence of cyclic nucleotides on junctional permeability in heart, *J. Mol. Cell Cardiol.,* 19, 763, 1987.

72. **Weidmann, S.,** The diffusion of radiopotassium across intercalated discs of mammalian cardiac muscle, *J. Physiol.,* 187, 323, 1966.

73. **Weingart, R.,** The permeability to tetraethylammonium ions of the surface membrane and the intercalated disks of the sheep and calf myocardium, *J. Physiol.,* 240, 741, 1974.

74. **Imanaga, I.,** Cell-to-cell coupling studied by diffusional methods in miocardial cells, *Experientia,* 43, 1080, 1987.

75. **De Mello, W. C.,** Cell coupling and healing-over in cardiac muscle, in *Physiology and Pathophysiology of the Heart,* Sperelakis, N., Ed., Kluver Academic Publishers, Boston, Massachusetts, 1989, 544.

76. **Brink, P. and Dewey, M.,** Nexal membrane permeability to anions, *J. Gen. Physiol.,* 72, 67, 1978.

77. **Loewenstein, W. R.,** Junctional intercellular communication: the cell-to-cell membrane channel, *Physiol. Rev.,* 61, 829, 1981.

78. **Einsenman, G.,** On the elementary atomic origin of equilibrium ionic specificity, in *Symposium on Membrane Transport and Metabolism,* Kleinzeller, A. and Kotyk, A., Eds., Academic Press, New York, 1961.

79. **Bennett, M. V. L.,** Gap junctional conductance: gating, in *Gap Junctions,* Hertzberg, E. and Johnson, R., Eds., Alan R. Liss, New York, 1988, 287.

80. **Saez, J. C., Spray, D. C., Nairn, A. C., Hertzberg, E. And Greengard, P.,** cAMP increases junctional conductance and stimulates phosphorylation of the 27-kDa principal gap junction polypeptide, *Proc. Natl. Acad. Sci. U.S.A.,* 83, 2473, 1986.

81. **Picolino, M., Neyton, J., Whitkovsky, P., and Gerschenfeld, H. H.,** GABA antagonists decrease junctional communication between L-horizontal cells of the retina, *Proc. Natl. Acad. Sci. U.S.A.,* 79, 3671, 1982.

82. **Cole, W. C.,** Gap Junctions in Uterine Smooth Muscle, Ph.D. thesis, McMaster University, Hamilton, Ontario, 1985.

83. **De Mello, W. C.,** interaction of cyclic AMP and Ca in the control of electrical coupling in heart fibers, *Biochem. Biophys. Acta,* 888, 71, 1986.

84. **Reuter, H. and Sholz, H.,** The regulation of the calcium conductance of cardiac muscle by adrenaline, *J. Physiol.,* 264, 49, 1977.

85. **Sperelakis, N. and Schneider, J. A.,** A metabolic control mechanism for calcium ion influx that may protect the ventricular myocardial cell, *Am. J. Cardiol.,* 37, 1079, 1976.

86. **Entman, M. L., Leveery, G. S., and Epstein, S. E.,** Mechanism of action of epinephrine and glucagon on the canine heart; evidence for increase in sarcotubular clacium stores mediated by cyclic 3'-5'-AMP, *Circ. Res.,* 25, 429, 1969.

87. **Kakiuchi, S., Yamazake, R., and Teshima, Y.,** Cyclic 3'-5'-nucleotide phosphodiesterase. IV. Two enzymes with different properties from brain, *Biochem. Biophys. Res. Commun.,* 42, 968, 1971.

88. **Cheung, W. Y.,** Calmodulin plays a pivotal role in cellular regulation, *Science,* 207, 19, 1980.

89. **Gabella, G. and Blundell, D.,** Nexuses between smooth muscle cells of the guinea pig ileum, *J. Cell Biol.,* 82, 239, 1979.

90. **Gabella, G.,** Structures of smooth muscles, in *Smooth Muscle,* Bulbring, E., Brading, A. F., Jones, A. W., and Tomita, T., Eds., Arnold, London, 1981, 1.

91. **Tomita, T.,** Electrical responses of smooth muscle to external stimulation in hypertonic solution, *J. Physiol.,* 183, 450, 1966.

92. **Daniel, E. E.,** Gap junctions in smooth muscle, in *Cell-to-Cell Communication,* De Mello, W. C., Ed., Plenum Press, New York, 1987.

93. **Engelmann, T. W.,** Vergleichende Untersuchungen zur Lehre von der Muskel- und Nervenel- ektricität, *Pflüg. Arch. Physiol.,* 15, 116, 1877.

94. **Weidmann, S.,** The electrical constants of Purkinje fibers, *J. Physiol.,* 118, 348, 1952.

95. **De Mello, W. C., Motta, G., and Chapeau, M.,** A study on the healing-over of myocardial cells of toads, *Circ. Res.,* 24, 475, 1969.

96. **De Mello, W. C.,** Cell-to-cell coupling assayed by means of electrical measurements, *Experientia,* 43, 1075, 1987.

97. **Fernandez, N. and De Mello, W C.,** Healing-over in smooth muscle, *The Physiologist,* 29, 4, 1986.

98. **Garfield, R. E., Sims, S., and Daniel, E. E.,** Gap junctions: their presence and necessity in myometrium during parturition, *Science,* 198, 958, 1977.

99. **Garfield, R. E., Sims, S. M., Kannan, M. S., and Daniel, E. E.,** The possible role of gap junctions in activation of the myometrium during parturition, *Am. J. Physiol.,* 235, C168, 1978.

100. **Garfield, R. E. and Hayashi, R. H.,** Appearance of gap junctions in the myometrium of women during labor, *Am. J. Obstet. Gynecol.,* 140, 254, 1981.

101. **Cole, W. C. and Garfield, R. E.,** Alterations in coupling in uterine smooth muscle, in *Gap Junctions,* Bennett, M. V. L. and Spray, D. C., Eds., Cold Spring Harbor Laboratory, Cold Spring Harbor, NY, 1985, 215.

102. **Dahl, G. and Berger, W.,** Nexus formation in the myometrium during parturition and induced by estrogen, *Cell Biol. Intern. Rep.,* 2, 381, 1980.

103. **Mackenzie, L. W., Puri, C. P., and Garfield, R. E.,** Effects of estradiol-17 β and prostaglandins on rat myometrial gap junctions, *Prostaglandins,* 26, 925, 1983.

104. **Mackenzie, L. W. and Garfield, R. E.,** Effect of tamoxifen citrate and cycloheximide on estradiol induction of rat myometrial gap junctions, *Can. J. Physiol. Pharmacol.,* 64, 703, 1986.

105. **Garfield, R. E., Merret, D., and Grover, A. K.,** Gap junction formation and regulation in myometrium, *Am. J. Physiol.,* 239, C217, 1980.

Chapter 2

MOLECULAR BIOLOGY OF MAMMALIAN GAP JUNCTIONS

Klaus Willecke and Otto Traub

TABLE OF CONTENTS

I. INTRODUCTION

Numerous recent reviews[1-7] and monographs[8-11] can guide the reader through almost all aspects of gap junction biology. This field entered the age of molecular genetics when the first cDNA codings for gap junction proteins were published 3 years ago. It is our intention to summarize the experimental results which led to this sequence information and to point out some of the conclusions which can be drawn from these studies. In this review, only recent and current references on gap junctions are discussed which pertain directly to the molecules from which gap junctions are formed. Since this area is changing very rapidly it is likely that some of the hypotheses and speculations of this review will already be verified or outdated when this book is published.

Gap junctions were discovered by electron microscopy as special morphological structures connecting membranes of apposed cells. Parallel studies demonstrated electrical conductance and (later) ion movement between apposed cells. These results have been attributed to the existence of cell-to-cell channels as communication units. The observations suggested that gap junctions consisted of clustered arrays of cell-to-cell channels, which can be demonstrated in nearly all cells of the mammalian body. This relatively old notion has recently been supported by direct experimental evidence demonstrating functional cell-to-cell channels reconstituted from purified proteins[12] or mRNA molecules[13] (see below). In the following section the different levels of gap junction gene expression will be discussed, beginning with gap junction proteins, since they were the first molecules studied following resolution with biochemical methods.

II. GAP JUNCTION PROTEINS

Since there is no apparent enzymatic activity associated with these cellular structures, gap junctions had to be purified from rat and mouse liver under conditions in which the preserved plaque structure could be used as a criterion for purification. In the presence of nonionic detergents most of the peripheral membranes dissolve. Gap junction plaques can be enriched by differential centrifugation and monitored by electron microscopy. Methods for purification of rat[14] and mouse liver plaques[15] have been independently worked out. Purified gap junction plaques were dissolved in sodium dodecyl sulfate (SDS) and subjected to polyacrylamide electrophoresis in the presence of SDS. A major protein band of 27 kDa apparent molecular weight, dimer proteins of about 47 kDa, and oligomers of higher molecular weight were found under these conditions. Presently it is not known whether protein bands of higher than 27 kDa apparent molecular weight represent protein complexes derived from the native channel structures or artificial aggregates of the monomeric proteins in the presence of SDS. Mouse liver plaques consist, in addition to the 26-kDa protein, of at least one other protein of 21-kDa apparent molecular weight, which was suggested to be a proteolytic degradation product of the 26-kDa protein[15] due to some common tryptic peptides. The 21-kDa protein is also present in rat liver plaques but only at about one tenth of the amount of the 27-kDa protein.[16]

Another method for isolation of gap junction plaques uses alkaline treatment of isolated membranes. Under these conditions a fraction of transmembrane proteins can be enriched and used for purification of gap junction plaques.[17] Again the 27-kDa protein was found to be the main component of this plaque preparation.

Parallel to biochemical studies of gap junction plaques in liver major intrinsic protein (MIP), so called because of its abundancy in lens, was purified from eye lenses from which the epithelial cell layer had been removed. This protein was later called MP26 because of its apparent molecular weight of 26 kDa and most recently called MP28,[18] since the theoretical molecular weight is 28,000 Da. MP26 is still assumed to be a structural component of gap junctions between lens fiber cells (for a review see Reference 18). Antibodies to the MP26 proteins do not cross-react with the 27-kDa protein from rat liver.[19,20]

We isolated and characterized antibodies to the mouse liver 26-kDa and 21-kDa protein.[21,22] These antibodies do not mutually cross-react, indicating that the 21-kDa protein could not be a degradation product of the 26-kDa protein.[21] We could not determine whether the 21-kDa protein was a structural component of the gap junction channel or an independent component of the plaque preparation from mouse liver.[23] Subsequently, using affinity-purified anti-21 kDa and electron microscopy, we found that these antibodies bind specifically to isolated hepatic gap junction plaques.[22] Furthermore, these antibodies recognized gap junctions on sections through mouse or rat liver. The antibody binding sites appeared as spots on contact membranes of apposed cells when analyzed by indirect immunofluorescence.[16] Each of these spots apparently represents a gap junction plaque, although the area of specific fluorescence is much larger than the real gap junction plaque due to amplification of the signal by multiple binding of fluorescencently labelled secondary antibodies.[24]

The antigenic determinant recognized by affinity-purified anti-hepatic 26 kDa was also found on sections through a number of other tissues, e.g., pancreas, kidney, uterus, small intestine, but not in heart and ovary.[25] On the other hand, affinity-purified 27-kDa antibodies, raised in sheep to rat liver gap junction plaques, reacted with all tissues which gave a positive immunoreaction with rabbit anti-26 kDa.[26] In addition, they also reacted with heart and ovary.[26] The resulting dilemma can now be explained. As is discussed below, different but related gap junction proteins are expressed in different mammalian tissues. Antibodies prepared in rabbits to the liver 26-kDa protein react only with this gap junction protein. However the sheep 27-kDa antibodies, which recognize the 27-kDa protein (now called connexin 32 [cx32] see below), also recognize the 43-kDa gap junction protein, expressed in myocardial and other tissues (now called connexin 43 [cx43]).[27] An apparent cross-reaction of rabbit anti-27 kDa with a 54-kDa protein band turned out to be due to a β-fibrinogen contamination of the antigen used for preparation of the antibodies.[28]

Some controversy has arisen in the past over the correct molecular weight of the major hepatic gap junction proteins based on measurements of the apparent molecular weight after SDS polyacrylamide gel electrophoresis. It is now clear that the designations 26-kDa, 27-kDa, and 28-kDa gap junction protein from rat or mouse liver all refer to the same protein of the same amino acid sequence. It has also been shown that variations of the electrophoretic conditions appear to yield an apparent molecular weight of about 32-kDa for the formerly called 26- to 28-kDa protein.[29] The real molecular weight of this protein based on cDNA sequences is 32,006 Da in rat[30] and mouse liver (see below).

Since the apparent molecular weight, based on SDS polyacrylamide gel electrophoresis, is rather variable, other parameters for characterization of gap junction proteins were urgently needed. One of these parameters is based on inhibition of dye transfer and electrical coupling between cells after microinjection of 27-kDa antibodies.[31] It is not clear how this inhibition occurs in molecular terms but it appears to be specific for anti-27 kDa and anti-21 kDa, at least in mouse hepatocytes.[22] Moreover, the N-terminal amino acid sequence of the 27-kDa rat liver protein was determined[32] and antibodies to an amino terminal oligopeptide were used for characterization of gap junction proteins.[33] More recently, the N-terminal amino acid sequence of the 21-kDa gap junction protein from rat and mouse liver was determined and shown to have high homology to that of the 26- to 28-kDa hepatic gap junction protein.[16]

Recently, a 70-kDa protein (occasionally found to be degraded to 64-kDa and 38-kDa proteins) was isolated from sheep eye lenses and shown to be in part homologous to the hepatic 27-kDa protein. Thus, it appears to be a member of the same gene family.[34] Using affinity-purified antibodies to the mouse liver 26-kDa and 21-kDa proteins, it was recently demonstrated that both antigenic determinants are located within the same hepatic gap junction plaques.[16,22] It is not yet known whether both proteins have the ability to participate in forming the same cell-to-cell channel or homomeric channels are clustered together within the same gap junction plaque.[22]

Although it was originally concluded, based on electrophoretic mobility, that myocardial tissue contains a 28-kDa gap junction protein,[36] careful inhibition of proteolytic activity in biochemical preparations of heart gap junctions led to identification of a 43-kDa gap junction protein.[37] Since this protein is homologous to the 21- and 26-kDa gap junction protein, based on comparison of cDNA sequences,[27] it is another member of the gap junction protein family (now called connexin family). Rat ventricular gap junction proteins appear to be linked by disulfide bridges and cannot be dissolved in 0.3% desoxycholate in contrast to rat atrial gap junctions.[38] The molecular basis for this observation is not known. In a recent study[39] we have analyzed the expression of different gap junction proteins in embryonic and adult rodent brain cells. Using affinity-purified antibodies to 27-kDa, 21-kDa, and 43-kDa proteins, we found that the 21-kDa is preferentially expressed before birth (in leptomeningeal cells, ependymal cells, and pinealocytes) and the 27-kDa preferentially after birth. The 43-kDa protein was detected in embryonic brain cells but it is highly expressed in adult brain cells (leptomeningeal cells, ependymal cells, astrocytes, pinealocytes). The 27-kDa is strongly expressed in adult oligodendrocytes. Our results demonstrate that the pattern of connexin expression changes in brain development and is altered in different adult brain cells. Previously it had been reported[40] that many tissues of the rat central nervous system reacted with sheep anti-27 kDa.[26]

It is not clear, at present, how many different proteins are true gap junction proteins in the sense that they are functional components of the gap junction channel. In addition to proteins of the connexin gene family, another protein of 16 kDa and 18 kDa apparent molecular weight has been found in gap junction plaques from vertebrate and invertebrate tissues, respectively.[41] The original isolation procedure for the 16-kDa mouse liver protein employed Triton X100 treatment of whole-tissue homogenates and trypsin treatment of extracted membranes.[42] When the addition of trypsin is omitted the 16-kDa protein is also found but the preparations of this protein are less pure.[43] Under their experimental conditions the authors did not detect any 27-kDa mouse liver protein in purified gap junction plaques.[41] The structure of the gap junction plaques prepared by this method appears to be very similar to that seen in conventional gap junction plaques using electron microscopy.[43] It was concluded that the 27-kDa protein can be completely removed from hepatic gap junction plaques and thus is either a contaminant or accessory gap junction protein, whereas the real gap junction protein has an apparent molecular weight of 16 kDa in vertebrates.[43]

We have confirmed that the 16-kDa protein is present as 3 to 20% of the total protein in gap junction plaques prepared from mouse liver.[22,35] Since most of our plaque preparations contained the 16-kDa protein as less than 5% of the total protein, the variable amounts of the 16-kDa could be due to the sensitivity of the 26 and the 21 kDa towards unwanted proteolytic degradation. The 16-kDa protein is more resistant towards degradation by trypsin. Antibodies that recognize the 16-kDa proteins on immunoblots of membrane protein and purified plaques have been described.[41,43] Unfortunately, the 16-kDa antibodies could not be used for immunofluorescence analysis of junctional and nonjunctional areas on hepatocytes.[41] Apparently the affinity of the available 16-kDa antibodies for their target is too low to allow this analysis. In contrast to different connexin proteins, the 16-kDa was found in gap junction plaques of all mammalian tissues which are known to contain significant amounts of gap junctions.[41] So far the limited but not published information concerning the amino acid sequence of the 16-kDa protein shows no homology to the connexin gene family.

It has been suggested that the 16 kDa mammalian protein could be a subunit of "pseudo" gap junctions, an array of gap junction-like particles seen on freeze-fractured electron micrographs.[44] However, the particles of pseudo gap junctions apparently do not penetrate the peripheral membrane, whereas the particles of the 16-kDa containing gap junction plaques do (J.D. Pitts, personal communication). Since the preparation of gap junction plaques, which were used for X-ray diffraction analysis, did not contain intact 27-kDa protein but only the 16-kDa protein, it was argued that the structure of the gap junction plaques derived from these studies is due to the 16-kDa subunit rather than due to the 27-kDa protein.[43] The total amino acid sequences derived from the corresponding cDNA sequences of the connexins clearly demonstrate helical regions of hydrophobic amino acids characteristic for transmembrane region of integral membrane proteins (see below). This analysis has not been possible for the 16-kDa since its amino acid sequence is not yet published.

The cornerstone of the argument that the 16-kDa protein is a gap junction protein appears to be the observation that preparations of Triton X100-purified plaques show the typical regular pattern of particles by electron microscopy and consist only of the 16-kDa protein.[43] We have repeated the Triton X100/trypsin method with mouse liver homogenates[42] and have found that the 27-kDa (cx32) protein is largely degraded after addition of trypsin (Traub, O., unpublished results). This is expected due to the many cleavage sites which are known to occur in the cytoplasmic region of the amino acid sequence of this protein.[30] Triton X100 does not dissolve purified gap junction plaques to single or aggregated hemichannels.[43] The transmembrane regions of the 27-kDa molecule are about 20 amino acids long (see Figure 1). We assume that preferentially the putative cytoplasmic parts of the 27-kDa protein are degraded in isolated gap junction plaques. Protein fragments of less than 10,000 kDa have been found in Triton X100/urea/trypsin-treated gap junction plaques which also contain the 16-kDa protein (Traub, O., unpublished results). This is in contrast to results of Finbow et al.,[43] who did not find protein fragments down to a molecular weight of 3,000 kDa or less. We think that proteolytic fragments of the 27-kDa protein contribute to maintenance of the pattern of regular subunits on specimens of Triton X100-treated gap junction plaques analyzed by electron microscopy.[43] The fact that these proteins or their fragments do not react with 26-kDa or 21-kDa antibodies could be explained by the assumption that the remaining fragments of these proteins carry very few if any immunoreactive sites for antibodies raised to the purified 26-kDa and 21-kDa proteins. Accordingly, our results argue against the conclusion[43] that the 26-kDa protein is an accessory gap junction protein whereas the 16-kDa protein is the only structural component of gap junctions.

It is possible that the 16-kDa protein participates in the structure of gap junctions. It may also contribute to the function of the cell-to-cell channel. The overwhelming experimental evidence supports the notion that 27 kDa (cx32) has the characteristics of the channel-forming protein within the gap junction plaque. Probably only sequence and reconstitution studies of the 16-kDa protein will allow clarification of its possible role for the structure and function of gap junctions.

III. CONNEXIN cDNAs AND CONNEXIN TOPOLOGY IN MEMBRANES

Using rabbit antibodies to the 27-kDa rat liver protein, the complete cDNA corresponding to this protein was isolated from a lambda gt11 cDNA expression library.[30] Moreover, another partial 27-kDa cDNA clone from rat liver and the complete cDNA corresponding to the human liver 27-kDa protein were described.[45] At the same time, a partial cDNA clone was isolated by its hybridization to synthetic oligonucleotides whose sequence had been derived from the N-terminal amino acid sequence of the rat liver 27-kDa protein.[46] Sequence analyses of the independently isolated cDNAs showed that they all coded for the expected N-terminal amino acid sequence of the 27-kDa protein. The derived total amino acid sequence was analyzed for its theoretical hydropathy.[47] This procedure yielded a topographical model of this protein with 4 transmembrane regions, each corresponding to about 20 relatively hydrophobic amino acids.[30,45] Support of the orientation, suggested by this model, came by analysis of protection towards action of proteases in preparations of purified rat liver plaques.[48] Recently, the immunoreactivity with antibodies to peptides of the 27-kDa protein was investigated[49] and supported the topological model derived from the hydrophobicity plots.[30,45] Figure 1 shows this general orientation. It should be pointed out that there is evidence that regions C and E (the carboxy terminus) are exposed to the cytoplasm. Furthermore, region B could be accessible from the cell surface when gap junction plaques were split by urea.[48] At this time there is no direct evidence supporting the notion that region A (the N-terminus) is exposed to the cytoplasm. The transmembrane region 3 has some polar residues suggesting that it forms an amphiphilic helix which could be oriented towards the hydrophilic pore of the putative membrane channel. Because of the hexagonal symmetry of the particles seen in hepatic gap junction plaques, six of the 27-kDa protein subunits may form the transmembrane channel as depicted in Figure 1 (lower part). It is important to note that this is so far a speculative model which needs further experimental proof before it can be accepted. For example, in the comparable case of the α-subunit of the nicotinic

27

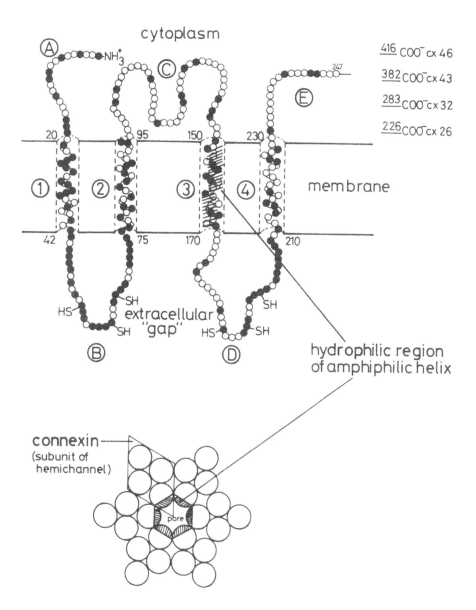

FIGURE 1. Hypothetical topological model of connexins as proposed in References 30, 45, 48, and 49 based on evalutions of hydropathy.[47] In the upper part of this figure the partial amino acid sequence of rat cx43 is schematically shown and compared with that one of rat cx32.[27,52] Black circles represent positions of identical amino acids and open circles represent positions of nonidentical or additional amino acids. The positions of amino acid homology are based on an optimized, computer-aided alignment of protein sequences as deduced from cDNAs for cx43 and cx32.[27,52] Optimized alignment means that a total of 21 amino acids of cx43 are missing in cx32 whereas two amino residues of cx32 are not represented in cx43. The encircled numbers and letters of the different connexin regions are shown as suggested.[51] In E the amino acid sequence ends at amino acid number 247 of cx43, which corresponds to amino acid number 228 of cx32. The total lengths of the different connexin sequences is indicated by the number of the corresponding carboxy terminal amino acid. Cx26 refers to connexin 26,[51] cx46 means connexin 46.[52] The transmembrane region 3 has an amphiphilic character. The lower part of this figure shows the hypothetical structure of the connexin hemichannel viewed perpendicularly to the plane of the membrane. The central pore could be formed from 6 amphiphilic regions each of which could be part of another connexin molecule. Each open circle represents a cross-section through a transmembrane region. Four of these transmembrane regions belong to one connexin molecule.

acetylcholine receptor, 2 amphiphilic transmembrane regions were originally deduced from analysis of the theoretical hydropathy. Only one, however, appears to be real, based on data of epitope mapping and site-directed mutagenesis (for review see Reference 50).

About a year ago, a new cDNA was isolated from rat heart cDNA which cross-hybridized under low stringency with the cDNA corresponding to rat liver 27-kDa protein. The new cDNA codes for a protein of 382 amino acids (43,036 Da).[27] It showed two regions of 58 and 42% homology compared to the cDNA probe corresponding to the rat liver 27-kDa protein. Thus both proteins are different members of the same gene family. The authors suggested the name connexin for proteins belonging to this gene family.[27] The different membranes are distinguished according to their theoretical molecular weight (in kDa) of the cDNA derived amino acid sequence. Thus the name cx32 was suggested for the 27-kDa protein. The 43-kDa gap junction protein was named cx43.

The hydropathy plot of the cx43 amino acid sequence yielded the same hypothetical topological model as derived for cx32.[27] The regions of relatively high homology between cx32 and cx43 include mainly sections 1, B, and 2 as well as 3, D, and 4 (Figure 1). Thus, the transmembrane regions (1, 2, 3, 4) and the presumptive extracellular domains (B, D) of the connexin molecules appear to be relatively conserved whereas the putative intracellular domains (A, C, E) are somewhat more variable.

The same conclusion is reached by comparison with two additional members of the connexin family, the connexin 26 (cx26),[51] which corresponds to the 21-kDa mouse and rat liver protein (see above), and connexin 46 (cx46), which has been found in rat lens fiber cells.[52] The cDNAs corresponding to each of these proteins have been sequenced and their preliminary characteristics have been published. Again the putative transmembrane regions and extracellular domains are highly conserved. However, the putative intracellular regions and in particular the length of the C-terminal amino acid sequence, which is believed to be exposed to the cytoplasmic domain of the gap junctions, are different between the different connexins. The position of three cysteine residues in each of the putative extracellular domains B and D are conserved among all four connexin sequences known to date.[51,52] The pattern of sequence conservation suggests that all connexin proteins can recognize each other via highly conserved extracellular sequence motifs. The conserved transmembrane regions may be necessary to stabilize the correct extracellular conformation of the connexin proteins. This conclusion suggests that different connexin proteins in hemichannels of different membranes have the ability to dock to each other and form complete cell-to-cell channels.

Several years ago it had been shown by dye transfer that cells of different tissues can communicate with each other.[53,54] The variability of the intracellular domains of the different connexin molecules suggests that these proteins may interact with different secondary molecules, which may recognize different motifs on different connexin molecules. One can speculate that gap junction channels consisting of different connexins in different cell types may be differently regulated. Whereas this speculation has to be proven by future studies, the distribution of the different connexin mRNA in different tissues can already be investigated using the connexin cDNA probes. A survey of the relative abundancy of connexin mRNAs shows that cx32 was found in liver, kidney, brain, and stomach,[28] whereas cx26 mRNA was detected in liver, kidney, and brain.[51] Cx43 mRNA was present in heart, lens epithelium, kidney, uterus, and ovary.[52] Cx46 mRNA has been found in lens fibers, lens epithelium, and to a lesser extent in kidney.[52] This shows that there are tissues like kidney which appear to contain at least four different connexins. It is not known, however, how many of these are expressed in the same cells of the kidney. In myocardial tissue, only cx43 mRNA has been demonstrated so far. In liver, cx32 and cx26 are expressed. In our laboratory we also found very small amounts of cx43 mRNA in rat liver and mouse embryonic hepatocytes (Stutenkemper, R. and Willecke, K., unpublished results). We do not yet know whether the hepatic cx43 mRNA is due to fibroblasts or due to hepatic precursor cells ("oval" cells). Several rat liver-derived cell lines (for example, BRL or RL cell lines) contain cx43 mRNA. The WB cell line, which is suggested to represent rat liver oval cells, contains cx43 and cx26 mRNAs but very little if any of cx32 mRNA. Furthermore, the mouse fibroblastoid L-cells as well as BALB/c 3T3 cell lines contain cx43 mRNA but harbor only traces of cx32 and cx26 mRNA. The BICR cell line, derived from a rat mammary tumor of Marshall rats, contains large amounts of cx43 mRNA. Mouse embryonic carcinoma F9 cells harbor cx43 mRNA (Stutenkemper, R. and Willecke, K., unpublished results). This suggests that cx43 may have had an ancestral role for mammalian connexins and that cx32 and cx26 may have been derived from cx43 by different deletions and other genetic alterations. All results are based on Northern blot hybridizations with the corresponding rat cDNAs.[27,30,46,51] So far there is no common pattern of expression of connexin mRNAs emerging. It is evident, however, that the former names of "liver" gap junction proteins and "heart" gap junction proteins are no longer acceptable. None of the different connexins appear to be expressed in only one tissue. It is possible that additional members of the connexin gene family exist and are expressed together with other connexins in certain tissues. Different connexin genes may be differently regulated during transcription. The connexin mRNAs in established cell lines can be quantitatively and qualitatively different from the original tissue. It is likely that the pattern of connexin expression will be regulated at the transcriptional level.

IV. CONNEXIN GENES AND TRANSCRIPTIONAL REGULATION

So far no detailed analysis of any connexin gene has been published. Thus, in discussing this aspect of connexin molecular biology one has to refer to preliminary results. The published cDNA sequences of rat connexins indicate that they are coded for by a family of different genes. At present it is not known whether or not alternative splicing occurs with certain connexin mRNAs. The rat cx32 gene[55] and the mouse cx32 do not have introns in the coding region (Kozjek, G. and Willecke, K., unpublished). The amino acid sequence of mouse cx32 appears to be identical to rat cx32. Twenty-one conservative base exchanges have been detected in the coding region and 31 base exchanges in the 3' noncoding region of the mouse cx32 gene (Kozjek, G. and Willecke, K., unpublished results). The results of Southern blot hybridizations suggest that both cx32 and cx26 are encoded by single genes in the rat genome.[28,51] Preliminary evidence indicates that the rat cx26 gene contains an intron in the 5' noncoding region and possibly another one in the coding region.[51] No genomic sequence data are available yet for the cx43 gene. Since the cx43 gene is highly expressed in the uterus[52] and since the morphologically recognizable gap junctions increase in size and number at term,[56] it is likely that there is an increase of the amount of cx43 transcript under these conditions. It is not yet known whether all or only certain of the connexin genes can be induced by steroid hormones. The inducibility by hormones or other stimuli may be one of the regulatory differences which led to the diversity of connexin genes during evolution.

So far there are only few examples for transcriptional regulation of cx32 mRNA. When primary mouse embryonic hepatocytes are taken in serum-free culture there is a drop in cx32 protein[22] and mRNA after 24 h.[46] After 3 days in culture, when the cells reach confluency, a 3- to 4-fold higher amount of cx32 mRNA is detected in these cells.[46] Thus, in embryonic mouse hepatocytes after 1 day in culture the rate of transcription or the stability of cx32 mRNA is much lower than in 3-day-old hepatocytes. The reason for this difference is not known. The rate of transcription and/or stability of cx32 mRNA may be different in proliferating and quiescent hepatocytes (cf. Reference 57) or may be dependent on cell contact. Recently, it has been reported[5] that the decrease and increase of cx32 after partial rat hepatectomy is due to a decrease and increase of cx32 mRNA. Furthermore, treatment of cultured hepatocytes from adult rat liver with heparins or proteoglycans increases cx32 mRNA, its protein, and its function as measured by dye transfer.[58] The connexin RNA expressed in fibroblastoid mouse L cells has been shown to be increased after treatment of the cells with cAMP.[59] Although the authors used cx32 cDNA as a probe under low stringency of hybridization, it is likely that their observation is due to induction of cx43 mRNA by cAMP.

V. RECONSTITUTION OF GAP JUNCTION CHANNELS

Ultimately, the solution of the discrepancies regarding the different proteins and their putative roles in gap junction structure and function should come from the successful reconstitution of cell-to-cell channels from their purified components. So far, two experimental systems have been successfully used to accomplish this task. Cx32 mRNA transcribed *in vitro* as well as cx32 cDNA under control of a highly inducible promotor have been injected into Xenopus oocytes.[13,59] Following injection of cx32 mRNA into paired oocytes, a 70-fold increase in conductivity can be measured. The low conductivity of endogenous channels can be distinguished from the induced activity by their sensitivity towards cytoplasmic acidification with CO_2.[59] Recently, the cx32 protein synthesized in Xenopus oocytes has been directly demonstrated by immunoprecipitation and immunoblot.[60] When pairing of injected Xenopus oocytes is delayed for 42 h a more rapid increase in conductivity is observed, suggesting the accumulation of subunit protein and rapid assembly after making contact.[59] The results indicate that the cx32 protein can form cell-to-cell gap junction channels. Endogenous Xenopus proteins may be required for the cell adhesion or cell recognition step, possibly preceding the cell-to-cell channel formation. Furthermore, it is possible that endogenous Xenopus proteins may contribute to the structure of the induced gap junction channels. Even if the liver cx32 protein is sufficient to form homomeric gap junction channels in the Xenopus oocytes, it is possible that heteromeric channels exist in liver. In spite of these questions the Xenopus oocyte system is, at the moment, the best understood reconstitution system of the gap junction channel. This system should allow measurement of the interaction between different connexins generated in different oocytes which are to be paired. It should also allow to follow the effects of site-specific mutagenesis of the connexin molecules on their biological function.

Ideally, one would like to reconstitute the gap junction channel in lipid bilayers in order to exclude any effect of endogenous proteins. It has been previously reported that gap junctional conductance was obtained when total proteins from purified gap junctions were incorporated into planar lipid bilayers.[12] Purified 27-kDa protein from rat liver plaques gave the same effect although the yield of functional cell-to-cell channels was very low under these conditions.[12] In both cases, the reconstituted channels were voltage dependent, a property that was not observed with gap junction channels in hepatocytes.[5,61] The experiments so far do not rule out the possibility that proteins "contaminating" the purified gap junction plaques can significantly contribute to the function of the cx32 protein. Only when reconstitution efficiencies with purified gap junction proteins comparable to the effects with total gap junction plaques are obtained can one evaluate the contribution of a single subunit protein. The experimental results thus far do not allow one to conclude that gap junction channels, induced in lipid bilayers, consist of hemichannels or complete channels residing in two contacting lipid films.

The MP26 protein, which does not show amino acid or nucleotide homology to the connexin family, has also been used for channel reconstitution in planar lipid bilayers[62] and in liposomes.[63] The results so far, however, are not very clear-cut and the system is difficult to control.[64] Crude MP26 antibodies do not unequivocally abolish the induced conductance.[64] Again, it is not clear whether hemichannels or complete channels are reconstituted in this system. If one accepts that functional voltage dependent MP26 channels have indeed been reconstituted then their properties are similar to the ones reconstituted from rat liver cx32 protein. Open hemichannels are unlikely to exist in living cells. Their experimental properties are difficult to predict. The published results of the reconstitutions do not allow one to decide whether or not functional complete gap junction channels can be formed from unrelated gap junction proteins.

VI. POSTTRANSLATIONAL MODIFICATIONS OF GAP JUNCTION PROTEINS

Using *mouse* embryonic hepatocytes, we showed that the cx32 protein is phosphorylated *in vivo*.[65] In pairs of adult *rat* hepatocytes, gap junctional conductance is increased by 30 to 80% in the presence of 1 mM solution of 8-bromo-cyclic AMP or dibutyryl cyclic AMP.[66] Forskolin, an activator of the catalytic subunit of cyclic AMP kinase, and glucagon stimulate gap junctional conductance by about 50%. The effect can be blocked by a protein inhibitor of cAMP dependent kinase.[66] The phosphorylated amino acid of the cx32 protein has been shown to be serine in rat[66] and mouse hepatocytes.[67] Only 2% of the incorporated $^{32}PO_4$ is bound to tyrosine, whereas 98% are found in serine residues of cx32 in mouse hepatocytes.[67] The putative cytoplasmic regions of rat and mouse cx32 (see Figure 1) contain several serine residues whose flanking amino acid sequence is related to consensus amino acid sequences for substrates of cAMP dependent protein kinase.[68] Sequencing studies of *in vivo* labeled connexins will have to show which of these serine residues is phosphorylated by cAMP-dependent kinase.

In addition to cAMP-dependent protein kinase, cx32 appears to be phosphorylated by calmodulin dependent kinase II[5] and by protein kinase C *in vitro*.[69] Chemical agents, which are known to elevate intracellular Ca^{2+} or activate protein kinase C, increase phosphorylation in intact rat cells and label the same tryptic peptides, which can be labeled after stimulation with cAMP.[5] The authors suggest convergence of these second messenger pathways since no uncoupling of rat hepatocytes with diacylglycerol and phorbol esters was observed.[5] It is possible that cx43, which is known to be expressed in fibroblasts, is phosphorylated in a different way than hepatic cx32. Recently, it has been reported that agents, which elevate intracellular cAMP in rat myocytes, increase junctional conductance whereas agents that increase intracellular cGMP depress junctional conductance.[70] In rat myocytes only cx43 is expressed. Phosphorylation of this protein has not yet been reported in intact cells.

In addition to its effect on serine phosphorylation of hepatic cx32, cAMP also increases the biosynthesis of the cx32 protein in cultured mouse hepatocytes probably at the transcriptional level.[67] It is likely that cAMP also increases the level of cx43 mRNA and -protein expressed in mouse L-cells.[59] We have studied phosphorylation of cx26 (i.e., the 21-kDa protein) in mouse embryonic hepatocytes. In cx26 we find less than 5% of the $^{32}PO_4$ incorporation as compared to cx32 expressed in the same cells.[22,35] No *in vitro* phosphorylation of cx26 with purified cAMP-dependent kinase was detected whereas cx32 was readily labeled under these conditions.[22] Thus, it seems that cx32 and cx26 are differently phosphorylated and probably differently regulated by phosphorylation in hepatocytes. It will be of interest to compare the functional consequences of connexin phosphorylation in different cells.

We have reported that cx32 and cx26 can be labeled in embryonic hepatocytes with radioactivity from ^3H palmitic or ^3H myristic acid in mouse embryonic hepatocytes.[35] Since the label is released after treatment with hydroxylamine it is likely that the long fatty acid residues are bound to cysteine residues. There are several cysteine residues in the amino acid sequence of mouse cx32. At present it is unknown whether palmitic and/or myristic residues are bound to the connexin proteins. Furthermore, the function of these putative long chain acyl residues for the transmembrane proteins is obscure. The need for additional membrane anchorage of connexins is difficult to visualize. Perhaps acyl residues are signals for protein degradation or recycling. The hepatic connexins do not appear to be connected to carbohydrate moieties.[3,14] This is unusual since protein domains expressed on the cell surface usually carry complex carbohydrates. Possibly, the presumptive connexin-connexin interaction, during docking of two hemichannels in contact membranes of different cells would be disturbed by the presence of carbohydrate chains on the extracellular domain of the connexin protein.

VII. CONCLUSIONS AND FUTURE DIRECTIONS

The exploration of gap junction structure has been aided enormously by application of the methods of molecular biology. The main result is the apparent diversity of gap junction proteins. Future studies will have to clarify whether this diversity is related to functional differences between the different connexins. Of primary importance is the solution to the question of additional (16-kDa protein?) or alternative (MP26?) gap junctional proteins in comparison with the connexin gene family. Are 6 connexin molecules indeed sufficient to form a complete functional gap junction channel in living cells? What cell recognition or cell adhesion proteins are required for the formation of gap junction channels? How is the assembly process of a gap junction plaque regulated which presumably starts with a single complete channel? Within a relatively short time the details of transcriptional regulation and control by hormones of connexin biosynthesis should be worked out. Future studies of functional reconstitution of gap junction channels in different expression systems are urgently needed to answer many questions about gap junction proteins. Their possible role in differentiation and growth control of mammalian cells can now be better experimentally approached. Recent results of cDNA isolation from Xenopus[71] will be useful for further studies of possible effects of gap junctional communication in amphibian development. The intercellular communication in early mouse development[72] has been investigated with rat liver cx32 antibodies.[73] In this context the blockage of gap junctions in Hydra by antibodies raised to the rat liver cx32[74] needs to be further investigated. Similarly as with Xenopus, connexin homologous cDNAs from Hydra will have to be cloned and used for studying morphogenesis in this organism. Furthermore, gap junction proteins (see Reference 75) and genes from Drosophila will have to be characterized before the full advantage of Drosophila molecular genetics can be exploited for this research field. The known methods of molecular biology guarantee that the structure of the gap junction forming molecules and presumably the structure of the cell to cell channels will be relatively well known within a few years. One can only hope, however, that detailed insights into the (multi-purpose?) function of these channels will also emerge.

ACKNOWLEDGMENTS

We thank our colleagues Drs. Rolf Dermietzel (Essen) and Bruce Nicholson (Buffalo) for communication of experimental results before publication. Work from our laboratory reported in this review was supported by grants from the Deutsche Forschungsgemeinschaft, the Dr. Mildred Scheel Stiftung-Deutsche Krebshilfe, and the Fonds der Chemischen Industrie.

REFERENCES

1. **Loewenstein, W. R.,** The cell-to-cell channel of gap junctions, *Cell*, 48, 725, 1987.
2. **Spray, D. C., Burt, J. M., White, R. L., and Wittenberg, B. A.,** Cardiac gap junctions: status and dynamics, in *The Conduction of the Cardiac Impulse,* Griles, W., Ed., Alan R. Liss, New York, 1987.
3. **Revel, J. P., Yancey, S. B., and Nicholson, B. J.,** The gap junction proteins, *Trends Biochem.,* 11, 375, 1986.
4. **Bennett, M. V. L. and Spray, D. C.,** Intercellular communication mediated by gap junctions can be controlled in many ways, in *Synaptic Function,* Edelman, G., Gall, E., and Cowan, W. M., Eds., John Wiley, New York, 1987, 109.

5. **Spray, D. C., Saez, H. C., and Hertzberg, E. L.,** Gap junctions between hepatocytes: structural and regulatory features, in *The Liver: Biology and Pathobiology,* Arias, I. M., Jakoby, W. B., Popper, H., Schachter, D., and Shafritz, D. A., Eds., Raven Press, New York, 1988, 851.

6. **Pitts, J. D. and Finbow, M. E.,** The gap junction, *J. Cell Sci.* (Suppl.) 4, 239, 1986.

7. **Warner, A.,** The gap junction, *J. Cell Sci.,* 89, 1, 1988.

8. **Hertzberg, E. L. and Johnson, R. G., Eds.,** *Gap Junctions,* Alan R. Liss, New York, 1988.

9. **Bock, G. and Clark, S., Eds.,** *Junctional Complexes of Epithelial Cells,* Ciba Found. Symp. 125, John Wiley & Sons, New York, 1987.

10. **De Mello, W. C., Ed.,** *Cell-to-Cell Communication,* Plenum Press, New York, 1987.

11. **Sperelakis, N. and Cole, W. C., Eds.,** *Cell Interactions and Gap Junctions,* CRC Press, Boca Raton, 1989.

12. **Young, J. D., Cohn, Z. A., and Gilula, N. B.,** Functional assembly of gap junction conductance in lipid bilayers: demonstration that the major 27 kDa protein forms the junctional channel, *Cell,* 48, 733, 1987.

13. **Dahl, G., Miller, T., Paul, D., Voellmy, R., and Werner, R.,** Expression of functional cell-cell channels from cloned rat liver gap junction complementary DNA, *Science,* 236, 1290, 1987.

14. **Hertzberg, E. L. and Gilula, N. B.,** Isolation and characterization of gap junctions from rat liver, *J. Biol.Chem.,* 254, 2138, 1979.

15. **Henderson, D., Eibl, H. J., and Weber, K.,** Structure and biochemistry of mouse hepatic gap junctions, *J. Mol. Biol.,* 132, 193, 1979.

16. **Nicholson, B., Dermietzel, R., Teplow, D., Traub, O., Willecke, K., and Revel, J. P.,** Hepatic gap junctions are comprised of two homologous proteins of MW 28000 and 21000, *Nature,* 329, 732, 1987.

17. **Hertzberg, E. L.,** A detergent-independent procedure for the isolation of gap junctions from rat liver, *J. Biol. Chem.,* 259, 9936, 1984.

18. **Johnson, R. G., Klukas, K. A., Houg, L. T., and Spray, D. C.,** Antibodies to MP28 are localized to lens junctions, alter intercellular permeability and demonstrate increased expression during development, in *Gap Junctions,* Hertzberg, E. L. and Johnson, R. G., Eds., Alan R. Liss, New York, 1988, 81.

19. **Zigler, T. S. and Horwitz, J.,** Immunochemical studies on the major intrinsic polypeptides from human lens membrane, *Invest. Ophthalmol. Vis. Sci.,* 21, 46, 1981.

20. **Hertzberg, E. L., Anderson, D. J., Friedlander, M., and Gilula, N. B.,** Comparative analysis of the major polypeptides from liver gap junctions and lens fiber junctions, *J. Cell Biol.,* 92, 53, 1982.

21. **Traub, O., Janssen-Timmen, U., Druege, P. M., Dermietzel, R., and Willecke, K.,** Immunological properties of gap junction protein from mouse liver, *J. Cell. Biochem.,* 19, 27, 1982.

22. **Traub, O., Look, J., Dermietzel, R., Brümmer, F., Hülser, D., and Willecke, K.,** Comparative characterization of the 21 kDa and 26 kDa gap junction proteins in murine liver, *J. Cell Biol.,* 108, 1039, 1989.

23. **Traub, O., Druege, P. M., and Willecke, K.,** Degradation and resynthesis of gap junction protein in plasma membranes of regenerating liver after partial hepatectomy or cholestasis, *Proc. Natl. Acad. Sci. U.S.A.,* 80, 755, 1983.

24. **Dermietzel, R., Yancey, B., Janssen-Timmen, U., Traub, O., Willecke, K., and Revel, J. P.,** Simultaneous light and electron microscopic observation of immunolabelled liver 27 kDa gap junction protein on ultra-thin cryosections, *J. Histochem. Cytochem.,* 35, 387, 1987.

25. **Dermietzel, R., Leibstein, A., Frixen, U., Drenckhahn, D., Traub, O., and Willecke, K.,** Gap junctions in several mouse tissues share antigenic homology with mouse liver gap junctions, *EMBO J.,* 3, 2261, 1984.

26. **Hertzberg, E. L. and Skibbens, R. V.,** A protein homologous to the 27000 Dalton liver gap junction protein is present in a wide variety of species and tissues, *Cell,* 39, 61, 1984.

27. **Beyer, E. C., Paul, D., and Goodenough, D.,** Connexin43: a protein from rat heart homologous to a gap junction protein from rat liver, *J. Cell Biol.,* 105, 2621, 1987.

28. **Gilula, N. B.,** Topology of gap junction protein and channel function, in *Junctional Complexes and Epithelial Cells,* Ciba Found. Symp., Bock, G. B. and Clark, S., Eds., John Wiley & Sons, Chichester, 1987, 128.

29. **Green, C. R., Harfst, E., Gourdie, R. G., and Severs, N. J.,** Analysis of the rat liver gap junction protein: classification of anomalies in its molecular size, *Proc. R. Soc. London,* 233, 165, 1988.

30. **Paul, D.,** Molecular cloning of cDNA for rat liver gap junction protein, *J. Cell Biol.,* 103, 123, 1986.

31. **Hertzberg, E. L., Spray, D. C., and Bennett, M. V. L.,** Reduction of gap junctional conductance by microinjection of antibodies against 27 kDa liver gap junctional polypeptide, *Proc. Natl. Acad. Sci. U.S.A.,* 82, 2142, 1985.

32. **Nicholson, B. J., Hunkapiller, M. W., Grim, L. B., Hood, L. E., and Revel, J. P.,** Rat liver gap junction protein: properties and partial sequence, *Proc. Natl., Acad. Sci. U.S.A.,* 78, 7594, 1981.

33. **Zervos, A. S., Hope, J., and Evans, W. H.,** Preparation of a gap junction fraction from uteri of pregnant rats. The 28 kDa polypeptides of uterus, liver and heart gap junctions are homologous, *J. Cell Biol.,* 101, 1363, 1985.

34. **Kistler, J., Christie, D., and Bullivant, S.,** Homologies between gap junction protein in lens, heart and liver, *Nature,* 331, 721, 1988.

35. **Willecke, K., Traub, O., Look, J., Stutenkemper, R., and Dermietzel, R.,** Different protein components contribute to the structure and function of hepatic gap junctions, in *Gap Junctions,* Hertzberg, E. L. and Johnson, R. G., Eds., Alan R. Liss, New York, 1988, 41.

36. **Gros, D. B., Nicholson, B. J., and Revel, J. P.,** Comparative analysis of the gap junction protein from rat heart and liver: is there a tissue specificity of gap junctions?, *Cell,* 35, 539, 1983.

37. **Manjunath, C. K., Goings, G. E., and Page, E.,** Cytoplasmic surface and intramembrane components of rat heart gap junctional protein, *Am. J. Physiol.,* 246, 4865, 1984.

38. **Manjunath, C. D. and Page, E.,** Structural differences between rat atrial and ventricular gap junctions, in *Gap Junctions,* Hertzberg, E. L. and Johnson, R. G., Eds., Alan R. Liss, New York, 1988, 69.

39. **Dermietzel, R., Traub, O., Hwang, T. K., Beyer, E., Spray, D. C., and Willecke, K.,** Differential expression of gap junction proteins in developing and mature brain tissues, submitted.

40. **Nagy, J. I., Yamamoto, T., Shiosaka, S., Dewar, K. M., Wittaker, M. E., and Hertzberg, E. L.,** Immunochemical localization of gap junction protein in rat CNS: a preliminary account, in *Gap Junctions,* Hertzberg, E. L. and Johnson, R. G., Eds., Alan R. Liss, New York, 1988, 375.

41. **Buultjens, T. E. J., Finbow, M. E., Lane, N. J., and Pitts, J. D.,** Tissue and species conservation of the vertebrate and arthropod forms of the low molecular weight (16-18000) proteins of gap junctions, *Cell Tissue Res.,* 251, 571, 1988.

42. **Finbow, M. E., Shuttleworth, J., Hamilton, A. E., and Pitts, J. D.,** Analysis of vertebrate gap junction protein, *EMBO J.,* 2, 1479, 1983.

43. **Finbow, M. E., Buultjens, T. E. J., John, S., Kam, E., Meagher, L., and Pitts, J. D.,** Molecular structure of the gap junction channel, in *Junctional Complexes of Epithelial Cells,* Ciba Found. Symp., Bock, G. and Clark, S., Eds., John Wiley & Sons, Chichester, 1987, 92.

44. **Berdan, R. C.,** Intercellular communication in arthropods, biophysical, ultralstructural and biochemical approaches, in *Cell-to-Cell Communication,* De Mello,W.C., Ed., Plenum Press, New York, 1987, 299.

45. **Kumar, N. and Gilula, N. B.,** Cloning and characterization of the human and rat liver cDNAs coding for a gap junction protein, *J. Cell Biol.,* 103, 767, 1986.

46. **Heynkes, R., Kozjek, G., Traub, O., and Willecke, K.,** Identification of rat liver cDNA and mRNA coding for the 28 kDa gap junction protein, *FEBS Lett.,* 205, 56, 1986.

47. **Kyte, J. and Doolittle, R. F.,** A simple method for displaying the hydropathic character of a protein, *J. Mol. Biol.,* 157, 105, 1982.

48. **Zimmer, D. B., Green, C. R., Evans, W. H., and Gilula, N. B.,** Topological analysis of the major protein in isolated intact rat liver gap junctions and gap junction-derived single membrane structures, *J. Biol. Chem.,* 262, 7751, 1987.

49. **Milks, L. C., Kumar, N. M., Hougton, R., Unwin, N., and Gilula, N. B.,** Topology of the 32 kD derived gap junction protein determined by site directed antibody locations, *EMBO J.,* 7, 2967, 1988.

50. **Maelicke, A.,** Structural similarities between ion chammel proteins, *Trends Biol. Sci.,* 13, 199, 1988.

51. **Nicholson, B. J. and Zhang, J.,** Multiple protein components in a single gap junction: cloning of a second hepatic gap junction protein (Mr 21000), in *Gap Junctions,* Hertzberg, E. L. and Johnson, R. G., Eds., Alan R. Liss, New York, 1988, 207.

52. **Beyer, E. C., Goodenough, D. A., and Paul, D. L.,** The connexins, a family of related gap junction proteins, in *Gap Junctions,* Hertzberg, E. L. and Johnson, R. G., Eds., Alan R. Liss, New York, 1988, 165.

53. **Michalke, W. and Loewenstein, W. R.,** Communication between cells of different types, *Nature,* 232, 121, 1971.

54. **Epstein, M. L. and Gilula, N. B.,** A study of communication specificity between cells in culture, *J. Cell Biol.,* 75, 769, 1977.

55. **Kumar, N. M.,** Characterization of gap junction genes and their products, in *Gap Junctions,* Hertzberg, E. L. and Johnson, R. G., Eds., Alan R. Liss, New York, 1988, 177.

56. **Dahl, G. and Berger, W.,** Nexus formation in the myometrium during parturition and induced by estrogen, *Cell Biol. Int. Rep.,* 2, 381, 1978.

57. **Dermietzel, R., Yancey, B., Traub, O., Willecke, K., and Revel, J. P.,** Major loss of the 28000 Dalton protein in gap junctions in proliferating hepatocytes, *J. Cell Biol.,* 105, 1925, 1987.

58. **Spray, D. C., Fujita, Y., Saez, J. C., Choi, H., Hertzberg, E. L., Rosenberg, L. C., and Reid, L. M.,** Glycosaminoglycans and proteoglycans induce gap junction synthesis and function in primary liver cultures, *J. Cell Biol.,* 105, 541, 1987.

59. **Dahl, G., Werner, R., and Levine, E.,** Paired oocytes: an expression system for cell-cell channels, in *Gap Junctions,* Hertzberg, E. L. and Johnson, R. G., Eds., Alan R. Liss, New York, 1988, 183.

60. **Swenson, K. I., Paul, D. L., and Goodenough, D. A.,** Biosynthesis of liver gap junction protein in Xenopus oocytes, *J. Cell Biol.,* 105, 263a, 1987.

61. **Bennett, M. V. L., Verselis, V., White, R. L., and Spray, D. C.,** Gap junctional conductance: gating, in *Gap Junctions,* Hertzberg, E. L. and Johnson, R. G., Eds., Alan R. Liss, New York, 1988, 287.

62. **Zampighi, G. A., Hall, J. A., and Kreman, D.,** Purified lens junctional protein forms channels in planar lipid films, *Proc. Natl. Acad., Sci. U.S.A.,* 82, 8468, 1985.

63. **Girsch, S. J. and Peracchia, C.,** Lens cell-to-cell channel proteins I. Self assembly into liposomes and permeability regulated by calmodulin, *J. Membr. Biol.,* 83, 217, 1985.

64. **Ehring, G. R., Zampighi, G. A., and Hall, J. E.,** Properties of MIP26 channels reconstituted into planar lipid bilayers, in *Gap Junctions,* Hertzberg, E. L. and Johnson, R. G., Eds., Alan R. Liss, New York, 1988, 335.

65. **Willecke, K., Traub, O., Janssen-Timmen, U., Frixen, U., Dermietzel, R., Leibstein, A., Paul, D., and Rabes, H.,** Immunochemical investigations of gap junction protein in different mammalian tissues, in *Gap Junctions,* Bennett, M. V. L. and Spray, D., Eds., Cold Spring Harbor Laboratory, Cold Spring Harbor, NY, 1985, 67.

66. **Saez, C., Spray, D. C., Nairn, A., Hertzberg, E. L., Greengard, P., and Bennett, M. V. L.,** cAMP increases junctional conductance and stimulates phosphorylation of the 27 kDa principal gap junction polypeptide, *Proc. Natl. Acad. Sci. U.S.A.,* 83, 2473, 1986.

67. **Traub, O., Look, J., Paul, D., and Willecke, K.,** Cyclic adenosine monophosphate stimulates biosynthesis and phosphorylation of the 26 kDa gap junction protein in cultured mouse hepatocytes, *Europ. J. Cell Biol.,* 43, 48, 1987.

68. **Edelman, A. M., Blumenthal, D. K., and Krebs, E. G.,** Protein serine/threonine kinases, *Annu. Rev. Biochem.,* 56, 567, 1987.

69. **Takeda, A., Hashimoto, E., and Shimazu, T.,** Phosphorylation of liver gap junction protein by protein kinase C, *FEBS Lett.,* 210, 169, 1987.

70. **Burt, J. M. and Spray, D. C.,** Inotropic agents modulate gap junctional conductance between cardiac myocytes, *Am. J.Physiol.,* 254, H1206, 1988.

71. **Gimlich, R. L., Kumar, N. M., and Gilula, N. B.,** Sequence and developmental expression of mRNA coding for a gap junction protein in Xenopus, *J. Cell Biol.,* 107, 1065, 1988.

72. **Kalimi, G. H. and Lo, C. W.,** Communication compartments in the gastrulating mouse embryo, *J. Cell Biol.,* 107, 241, 1988.

73. **Lee, S., Gilula, N. B., and Warner, A. E.,** Gap junctional communication and compaction during preimplantation stages of mouse development, *Cell,* 51, 851, 1987.

74. **Frazer, S. E., Green, C. R., Bode, H. R., and Gilula, N. B.,** Selective disruption of gap junctional communication interferes with a patterning process in hydra, *Science,* 237, 49, 1987.

75. **Frazer, S. E. and Bryant, P. J.,** Patterns of dye coupling in the imaginal wing disc, *Nature,* 317, 533, 1985.

Chapter 3

ELECTRICAL COUPLING BETWEEN MUSCLE CELLS IN CULTURE

Ida Chow

TABLE OF CONTENTS

I. INTRODUCTION

The presence of direct intercellular communication via electrical signals was first demonstrated by Furshpan and Potter in the crayfish.[20] The presumptive structure responsible for this electrical coupling, the gap junction, was first described by Robertson, also between giant axons in the crayfish.[45] The existence of this direct pathway between the cytosols of adjacent cells has been further confirmed by the passage of fluorescent dyes,[26,27,57] metabolites,[41,52] or electrical currents[4,20,31,53] between confluent cells. These observations suggest that molecules up to about 1000 Da in molecular weight can spread through adjacent cells via these intercellular channels, without passage through the extracellular space.

Gap junctions are regions of membrane in close apposition (2 to 4 nm) with channel structures which form a continuous passageway between the cytoplasms of adjacent cells. Each cell contributes one half of the channel (the hemichannel) which, as expected, extends through the entire thickness of the cell membrane. Most of the gap junctions found in adult epithelia, liver, nervous tissue, smooth and cardiac muscles are formed by plaques of channels in hexagonal arrays with 8- to 9-nm center-to-center spacings. Each channel is a hollow hydrophilic cylinder about 6 nm in diameter with a 1- to 2-nm pore (see recent review by Zampighi[66]).

Most of the electrophysiological studies using intracellular recording suggest that these junctional channels behave similarly to other membrane channels in having some sort of gating properties (see review by Loewenstein,[31] Spray and Bennett,[54] De Mello,[16] and Bennett et al.[5]), such that the opening of the channels can be controlled by variations in pH,[56,58,59] intracellular calcium concentration,[37,39,44,46,47] calmodulin,[40] cross-junctional potentials,[20,55,56] or some other factors.[3,13,25,30,35]

With the introduction of the patch clamp technique[22] it is now possible to measure directly the current that flows through a single gap junction channel because of the high resolution of this recording method. This technique is based on the very tight seal between the patch pipette and the cell membrane which provides a seal resistance two to three orders of magnitude higher than that of the pipette. In turn, this pipette resistance is many times smaller than that of the cell membrane resistance, so that very small currents going through single membrane channels can be detected. For measurements of single gap junction channel conductance the cells are held under whole-cell patch clamp mode so that the transjunctional voltage can be held constant and currents that flow through the junction can be recorded by the patch pipette (see details below). To date, the reported mean values for single gap junction channel conductances are 120 picoSiemens (pS) for lacrimal cells,[35] 165 pS for chick embryonic heart cells,[60] 114 pS for frog myotomal muscle cells,[65] and 53 pS for neonatal rat heart cells.[7] The average time a channel remains open is long: 2 s for lacrimal cell channels, 332 ms for chick heart cells, 31 s for myotomal muscle cells, and 0.95 s for neonatal rat heart cells. This long open time may have significant physiological consequences, since the amount of substances that can pass from one cell to the other through a lengthy period may be large even if the total number of open channels is small (see below).

The diffusion rate of molecules crossing the gap junctions can be estimated by timing the transfer of fluorescent dye which is injected into one cell and later appears in the neighboring cells (see Reference 49). The most common dyes used are Lucifer Yellow (M_r 457) and carboxyfluorescein (M_r 376); the former is usually injected into the cell and the latter applied to the bathing solution in a membrane permeable form. Due to inherent limitations in the detection of small amounts of fluorescent dye, one has to be cautious in interpreting the absence of dye passage in conditions of low coupling as complete lack of coupling. Presently, this drawback can be partly remedied with the use of image intensifying procedures. For molecules of the size of approximately 400 Da a diffusion coefficient of 1 to 4×10^{-7} cm^2/s was obtained by Safranyos and Caveney.[49]

Much of what is presently known on electrical coupling comes from studies of cells that are already coupled. Only a few studies have dealt with the process of coupling formation. This is mainly due to the lack of a preparation where the exact timing of onset of coupling can be determined. In addition, very little is known about electrical coupling between skeletal muscle cells and what regulates this particular form of intercellular communication.

Most of this chapter will summarize the results obtained from studies of electrical coupling in myotomal muscle cells isolated from embryos of the South African clawed frog *Xenopus laevis*. These striated muscle cells were kept in culture and were amenable to manipulation, allowing precise timing of cell-cell contact and subsequent onset of electrical coupling. Since coupling was frequently detected between cells that were found already in contact in these cultures, some of the properties of these developed junctions could be studied in the same preparation. Furthermore, this preparation seems to be ideal to address the questions on modulation and regulation of electrical coupling because of precise temporal and spatial control, as well as direct access to the interior of the cells by using the patch pipette under whole-cell mode.

II. COUPLING BETWEEN MUSCLE CELLS *IN VIVO*

The presence of gap junctions between differentiating muscle cells has been reported in developing tadpoles,[6] neonatal rats,[50] and cultured myoblasts[29,43] and their possible role in cell fusion was suggested,[29,43,50] although not yet directly proven.

In the developing *Xenopus laevis,* electrical coupling was detected between myotomal muscle cells both within and between the somites.[6] It gradually disappeared 1 or 2 days later, after neuromuscular transmission was established.[2] Since innervation of the myotomes occurs craniocaudally[9] and coordinated body movements take place before all the myotomes are innervated, this behavior is likely to result from intercellular propagation of membrane depolarization via gap junctions throughout the larval somites. Armstrong et al.[2] found that synaptic transmission block (by tricaine and binding of the acetylcholine [ACh] receptors with α-bungarotoxin) before the onset of uncoupling kept the myotomal muscle cells coupled. Apparently, the uncoupling depends to some degree on the activation of ACh receptors by neural transmission. Suppression of muscle cell membrane electrical activity maintains coupling between these cells.

On the other hand, less is known about the formation of electrical coupling in these myotomal muscle cells *in vivo*. The mesodermal cells, presumptive precursors of myotomal cells, are initially coupled; they uncouple during segmentation and become coupled again after somites are formed.[6] What regulates the different coupling formation processes? How many channels open when a new junction forms? Although many hypotheses have been suggested for coupling formation (see reviews by Gilula,[21] and Loewenstein[31]), little can be done using the intact embryo due to difficulties for the direct access to the cells. Since the cells are all aggregated and forming tissues, study on single cell-cell contacts cannot be carried out. In addition, precise timing for the onset of coupling in the developing embryo is just about impossible. As a result, one would have to resort to the use of cells in culture to answer these questions because cell density and cell-cell contact can be experimentally controlled.

III. COUPLING FORMATION BETWEEN
CELLS IN CULTURE

Most of the studies presently available report properties of gap junctions which are already present between different cell types, such as blastomeres, salivary gland cells, hepatocytes, cell lines, neurons, and cardiac and smooth muscle cells. Less is known about what is involved in the process of the junction formation except that coupling develops very quickly, in the order of a few minutes, in cell lines kept in culture[24,36,51] and between embryonic cells.[10,23,32] Inhibition of protein synthesis did not prevent gap junction formation in mouse blastomeres,[33] in epithelial cells,[42] and Novikoff cells, where lack of ATP also had no effect.[18] Treatment with antibodies to neural cell adhesion cell molecules caused a decline in the junctional transfer of fluorescent dye between chick neuroectodermal cells in culture.[28] Thus, it seems that gap junction precursors are readily available in the cell membrane and that low-energy lateral diffusion is sufficient for the precursors to pair up and form a junctional channel as long as close contact takes place at the cell-cell contact site.

Another question to be asked is the number of junctional channels that open at the onset of coupling. Do all the channels in a plaque open at once, or only one channel open at a time? Loewenstein et al.,[32] using a phase-sensitive recording method, described step-like increases of junctional conductance during coupling formation in *Xenopus* blastomeres, which they interpreted as successive junctional channel openings. However, this indirect method did not provide information on the exact number of open channels.

Microinjection of mRNA obtained from estrogen-stimulated rat myometrium into *Xenopus laevis* oocytes allowed formation of electrical coupling between paired oocytes after a delay of about 4 h.[63] The extent of coupling increased rapidly until about 12 hours and it was more gradual afterward. A similar time course was found between oocytes paired up just prior to mRNA injection, 12 h after injection, or uninjected oocytes which had endogenous coupling capability. The extent of coupling between uninjected pairs was, however, about five-fold lower than that between injected pairs after 12 h of contact. Since delayed pairing of oocytes after mRNA injection did not accelerate coupling formation, the authors suggested that the rate-limiting steps could be those following precursor insertion into the membrane and leading to the channel formation when the number of inserted precursors is low. If this was the case, some other additional factor(s) must play a role because in other coupling competent cell types direct contact between isolated cells leads to coupling formation within minutes (see above).

Rash and Fambrough[43] monitored continuously intimately associated pairs of cultured chick and rat myogenic cells (myotubes and/or myoblasts) for the appearance of electrical coupling as an indication of onset of cell fusion. They iontophoretically applied ACh onto the surface of one of the cells of the pair and recorded intracellularly from the other cell. This gave them the time course for establishment of passage of the ACh potential from one cell to the other. Subsequent ultrastructural analysis of serial sections of the cell pair provided morphological signs of cell fusion. They found that in some cases an initially weakly coupled pair would suddenly increase its coupling efficiency, and in less than 3 min of this jump the myotubes were completely fused. In situations where coupling efficiency between the two cells remained low, no fusion was observed up to 22 min after formation of coupling. The authors proposed that the gap junctions may mediate an equilibrium of ions or an exchange of organic molecules between the cells, or the junctions may maintain the outer surfaces of the apposed membranes together, as a preparatory step for myogenic cell fusion which takes place very quickly. Whether any or all of these alternatives is regulating myotube fusion remains to be elucidated.

IV. COUPLING BETWEEN *XENOPUS* MYOTOMAL MUSCLE CELLS IN CULTURE

A. COUPLING FORMATION AS DETECTED BY INTRACELLULAR RECORDING

The time course and the factors that may regulate electrical coupling formation were investigated using cultures of *Xenopus* myotomal muscle cells where the extent of coupling was monitored with intracellular microelectrodes.[10]

Somites were isolated from young *Xenopus* embryos and myotomal cells, mainly muscle cells, were dissociated and plated on clean, uncoated glass microscope slides. After 1, 2 or 5 days of culture, both spindle shape and spherical muscle cells (myoballs) were found, the latter due to the lack of a strongly adhesive surface. Isolated myoballs were detached from the slide surface using micropipettes and then translocated into contact with one another. The extent of electrical coupling after different contact periods was measured by monitoring the electrotonic spread of membrane potential changes from one cell into another. Potential changes were induced either by intracellular current injections, or by extracellular iontophoretic application of ACh on the surface of one of the cells. In these muscle cells the ACh receptors have a widespread distribution over the cell membrane and iontophoretic application of this neurotransmitter on a cell's surface induces depolarization which can propagate into the neighboring cell if gap junctions are present. To establish the exact onset time of coupling after initial cell-cell contact, an isolated myoball was impaled with an intracellular microelectrode, detached from the surface and pushed against another myoball, while the resting membrane potential was monitored continuously with the microelectrode. Figure 1 is a phase contrast photomicrograph showing the manipulation of 2 myoballs into contact.

Coupling was undetectable immediately after cell-cell contact. In 1-day old cultures, electrotonic spread of ACh-induced depolarization from one cell to the other began as early as 2 min after contact was made and increased rapidly with time, reaching a plateau within 20 min. The delay for coupling onset varied from cell pair to cell pair and after 30 min contact 66% of the cell pairs were found to be coupled. In contrast, only 24% of the cell pairs in 2-day old cultures and none of the cell pairs in 5-day old cultures was electrically coupled after 30 min contact. The number of

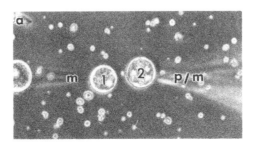

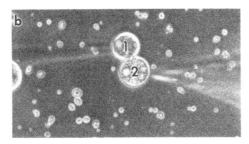

FIGURE 1.   Phase contrast photomicrographs showing manipulation of the *Xenopus* myotomal muscle cells in culture. (a) An isolated myoball (1) is impaled with an intracellular microelectrode (m) and detached from the glass culture slide surface, and another isolated myoball (2) is either placed under a whole-cell voltage clamp with a patch electrode (p), or has ACh iontophoretically applied onto its surface with a microelectrode (m). (b) Cell 1 is pushed into contact with cell 2 and the length of cell-cell contact area is about a cell radius. Diameter of these cells ranges between 25 to 30 μm.

coupled pairs did increase with contact time, even in the older cultures, so that after 12 h contact coupling was detected in over 70% of the cell pairs. Interestingly, the extent of coupling, as expressed in coupling ratio (propagated ACh potential/depolarization induced by applied ACh), was similar in all age groups and after 0.5, 1, 2 and 12 h contact, with an average value of 0.25, suggesting some sort of upper limit in the total number of gap junction channels to be formed and kept open.

The initial delay for coupling formation in 2- and 5-day-old cultures was probably caused by the appearance of extramembranous constraints for immediate close contact between the cell membranes upon contact and/or lateral mobility of channel precursors to the contact site. Pretreatment of the cultures with colchicine, a mixture of metabolic inhibitors (dinitrophenol, sodium azide and sodium fluoride), calcium-magnesium-free saline solution with ethylenediaminetetraacetic acid (EDTA) or trypsin increased significantly the rate of coupling formation to a level close to that of younger cells. Presumably these drugs affect the integrity of the extramembranous structures. Trypsin and calcium-magnesium-free saline may remove extracellular matrix,[1,8,12] whereas metabolic inhibitors and colchicine may disrupt cytoskeletal structures.[38,61,64] In fact, the development of cell coat and cytoplasmic filamentous material has been described in these cells.[62] Again, the average coupling ratio in these pretreated cell pairs was similar to the younger cell pairs' value.

These studies indicate that the junctional channel precursors are readily available

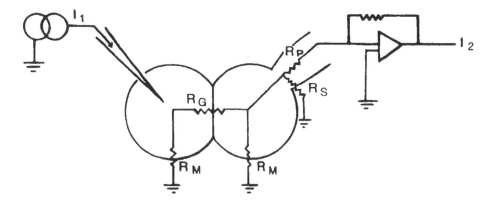

FIGURE 2. Equivalent circuit diagram showing the recording system using both intracellular microelectrode (left cell) and patch clamp in whole-cell voltage clamp mode (right cell). Current pulses (I_1) are applied into cell 1 through the microelectrode, a fraction of this current passes through gap junction channels into cell 2 and is recorded by the patch pipette (I_2). R_g = gap junctional resistance, R_m = nonjunctional cell membrane resistance (75 MΩ), R_p = patch pipette access resistance (5 MΩ), R_s = seal resistance (10 GΩ), I_1 = 500 pA, I_2 = 3.75 pA, R_g = 500pA/3.75pA $\times$ 75MΩ = 10,000 MΩ, and junctional conductance = $1/R_g$ = 100 pS.

in the cell membrane and the gap junctions form rapidly after the cells come into contact, establishing direct communication between the cytosols. In addition, as long as physical constraints are removed, the channel precursors can quickly diffuse in the plane of the cell membrane to the contact site, match up, and assemble into intercellular channels, establishing electrical coupling. It is interesting that the same extent of coupling can be reached in both the young and the older cells, even though the time course was different. This suggests that the extent of coupling is somehow regulated in these cells.

B. OPENING OF SINGLE JUNCTIONAL CHANNELS DURING COUPLING FORMATION

In order to investigate in greater detail the process of coupling formation, specifically to determine the number of junctional channels that open at the onset of coupling, Chow and Young[11] used a combination of whole-cell patch and intracellular recording techniques to record coupling formation at the single-channel level in the *Xenopus* myotomal muscle cells.

Figure 2 represents the equivalent circuit of the combined whole-cell patch clamp (right cell) and intracellular microelectrode (left cell) recording techniques used in that study.[11] The microelectrode was used to manipulate one of the muscle cells into contact with the other (since this manipulation is almost impossible with the patch pipette) and to inject current pulses once contact was established. Instead of constant current, current pulses were applied to assure that measured currents were in fact crossing the junction. Any current which passes between the two cells must be in synchrony with the injected current pulses. The very high seal resistance (R_s) between the patch pipette and the cell membrane provides very high resolution for the measurement of the currents that flow through the gap junction from the impaled to the patched cell. Since R_s is two to three orders of magnitude higher than that of the pipette (R_p) and R_p is ten times smaller than the cell membrane resistance (R_m), the values of the current flowing through the gap junction can be considered nearly equivalent to those recorded by the patch pipette. Thus, the current fluctuations recorded with the patch pipette that occur synchronously with the injection of constant amplitude current pulses from the intracellular microelectrode indicate the opening of gap junctional channels between the two cells.

An isolated myoball was impaled with a microelectrode, detached from the slide surface, and translocated into contact with another isolated myoball firmly attached on the glass surface (Figure 1). The second cell was then placed under whole-cell voltage clamp with a patch pipette and held at the cell resting potential (−70 to −100 mV). Hyperpolarizing current pulses of constant amplitude (500 pA) and duration (300 ms) were injected through the microelectrode into the impaled myoball. Any current passing through the gap junction channels into the second cell was recorded by the patch clamp amplifier in the whole-cell configuration. For an average channel current of 3.7 pA (see below) and an average cell membrane resistance of 75 MΩ, an estimated value of 100 pS was obtained for a single gap junctional channel conductance during the onset of coupling formation.

The time for the first appearance of an open channel after cell-cell contact was highly variable, ranging from 30 s to 3 min. The channels remained open for long periods of time, in the range of 10 to 40 s, and infrequently closed for short periods, with an average closed time of 0.8 s. Channels opened and closed with fast transition times, less than 1 ms. The initial amplitude of the current recorded in the patched cell remained constant and, after various time intervals, step increases of integral multiples of a smallest level (average 3.7 pA) in the amplitude were observed. These results indicate that when the gap junction was formed only one channel opened and subsequent channels appeared one at a time after lag times varying from 23 s to 3 min. Figure 3 shows examples of diagrams of times of new junctional addition at two newly formed gap junctions.

C. SINGLE JUNCTIONAL CHANNEL PROPERTIES

Junctional channel properties were also studied on cell pairs found in contact in culture and which had already established stable coupling before the experiments.[65] This could be done using double-patch technique (whole-cell voltage clamp mode in each cell) because no manipulation of the cells was necessary to form junctions. A linear relation was found between junctional current and junctional voltage. The straight line passed through zero, indicating that these channels are not voltage-dependent between −40 and +50 mV transjunctional voltage. The channels remained open for long periods of time, with a mean open time value of 31 ± 9 s ($\pm$ SE), and infrequently closed for short periods, with mean closed time value of 0.8 ± 0.2 s, values similar to those of newly formed channels. Similarly, the opening and closing of these established channels had a very fast transition time, less than 1 ms, with no detectable intermediary steps. Single gap junction channel conductance varied from 85 to 140 pS, with a mean value of 114 ± 10 pS ($\pm$ SE). These values are similar to the one estimated for the newly formed junctional channel using one patch electrode and one intracellular microelectrode (100 pS), thus suggesting that forming gap junction channels are similar to established gap junction channels. It is also interesting to note the range of conductance values, which may suggest differences in the diameter of the channel, or

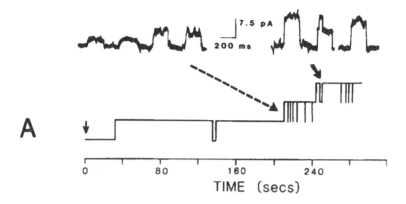

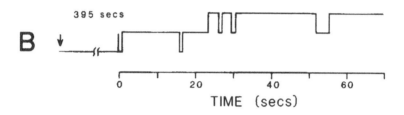

FIGURE 3. Diagrams showing formation of new gap junctional channels in 2 experiments. The vertical arrows point to the time of initial cell-cell contact, at the baseline, without coupling, i.e., no passage of currents. Channel opening amplitudes are in arbitrary units. (A) The entire kinetic history of one experiment with tracings of current pulses recorded at two selected times after cell-cell contact. There was a delay of 30 s after cell contact before the first channel opened. The channel remained open for long periods of time (mean open time 30.9 ± 8.9 s [$\pm$ SE of mean]) and closed for brief periods of time (0.8 ± 0.2 s). The first inset shows the transition between one and two open channels (dotted line). The second inset shows the temporary closure from 3 open to 2 open channels, and an opening back to 3 open channels 2.2 s later (solid line). (B) In this second experiment the first channel opened after 395 s of cell contact. The estimated mean open time is 15.0 ± 5.6 s, and the estimated mean closed time is 0.8 ± 0.3 s. (From Chow, I. and Young, S. H., *Devel. Biol.*, 122, 332, 1987. With permission.)

that different numbers of channel subunits were assembled. Since no intermediary steps were observed, it is unlikely that the different conductance values could be due to different opening stages. It is conceivable, however, that the smaller values are due to incomplete opening of one of the hemichannels. Incidentally, this wide range in channel conductance has also been reported in other tissues and the underlying mechanism needs to be investigated.

With the average coupling ratio value of 0.25 obtained from the intracellular studies[10] the entire gap junction between two average myoballs would have a total conductance of about 3300 pS, which now translates into a total number of 29 open channels. Although this absolute number of channels is small, they have high conductance and long open time, which allows the passage of significant amount of regulatory molecules between neighboring cells. This may be an important asset to these developing muscle cells because instruction molecules may flow from one cell to the other via the gap junction channels so that the cells may uncouple later on, once innervation is completed.

D. MAINTENANCE OF ELECTRICAL COUPLING

When at least one of the two cells of the pair is held under whole-cell voltage clamp with a patch pipette filled with an internal saline solution containing salts only, the cell pair will become completely uncoupled.[65] Uncoupling occurred even when internal Ca^{++} and pH were heavily buffered (11 mM [ethylene-bis(oxyethylenenitrilo)] tetraacetic acid, 1 mM $CaCl_2$, 20 mM Hepes, pH 7.8). The average time for complete uncoupling was 16 min. This time course is much slower than the time required for exchange of small ions, such as K^+, between the interior of the cell and the perfusion fluid in the pipette, which ranges between 2 and 4 min in these cells. This suggests that the modulating factor(s) that was being washed out by perfusion of the cytoplasm did not exert its effects immediately upon the channels. Alternatively, a larger size of the modulatory molecule or binding of this molecule to other cytoplasmic structures may have resulted in slower wash-out, thus the longer time for uncoupling to take place.

When 1 mM cyclic AMP was added into the internal perfusion solution in the patch pipette the uncoupling process was prevented. The exact mechanism by which this nucleotide maintains coupling is not known yet, although increase in coupling has been reported in cardiac cells,[14,15,17] mammalian cell lines,[19] and rat hepatocytes.[48] Interestingly, addition of cAMP into the patch pipette did not promote formation of coupling, indicating that this molecule exerts different effects on the processes of formation and maintenance of electrical coupling in myotomal muscle cells. Furthermore, continuous pulsing of current and recording of the cell pair with intracellular microelectrodes did not uncouple the pair, supporting the notion that some soluble factor was being removed from the cytoplasm by perfusion of the patch pipette solution.

E. DYE COUPLING

Since only an average of 30 open channels are present at the peak of coupling (see above), it would be interesting to find out the time course for detection of dye passage from one cell to the other. This experiment would also tell us whether the long open time of these channels does in fact allow sufficient flow of molecules of a weight similar to that of the fluorescent dye from one cell to the other within useful time periods. Lucifer Yellow CH (0.3%) was perfused with a patch pipette into one muscle cell within a cluster of muscle cells, no stain was seen in the neighboring cells immediately and up to 30 min later. A small amount of fluorescence was detected 2 h later in one of the neighboring cells[65] (Figure 4). Although this delay seems to be long, it is not necessarily so in the life of a developing *Xenopus* larva, where coupling between the myotomal muscle cells remains for a few days, gradually disappearing after innervation is completed. This intercellular communication may be crucial for the passage of putative messengers resulting from innervation of the muscle cells.

V. CONCLUSIONS

The *Xenopus* myotomal muscle cells are a suitable system for the study of electrical coupling because they can be manipulated at the investigator's will, thus providing

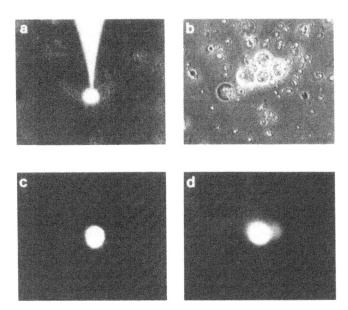

FIGURE 4. Fluorescent (a, c, and d) and phase contrast (b) photomicrographs showing dye passage between *Xenopus* myotomal muscle cells in culture. (a) Lucifer Yellow was perfused into one muscle cell within a cluster of muscle cells with a patch pipette. (b) Phase contrast of the muscle cluster. (c) Thirty minutes after introduction of the dye, no noticeable fluorescence was found in the neighboring cells. (d) Two hours later, fluorescence was seen in one of the neighboring cells.

very high temporal resolution. Since they can be readily patched and held in the whole-cell mode, very high resolution recording can be done, as well as access to the interior of the cell's cytoplasm for introduction of potential modulators or regulatory factors.

ACKNOWLEDGMENTS

I thank Dr. S. H. Young for critical comments on the manuscript and for providing Figure 4. I thank Dr. A. D. Grinnell for continuous support. Support from grants from the Muscular Dystrophy Association (to author) and the National Institutes of Health (to Dr. A. D. Grinnell) is also gratefully acknowledged.

REFERENCES

1. **Anghileri, L. J. and Dermietzel, C.,** Cell coat in tumor cells — effects of trypsin and EDTA: a biochemical and morphological study, *Oncology,* 33, 17, 1976.
2. **Armstrong, D. L., Turin, L., and Warner, A. E.,** Muscle activity and the loss of electrical coupling between striated muscle cells in *Xenopus* embryos, *J. Neurosci.,* 3, 1414, 1983.

3. **Azarnia, R., Dahl, G., and Loewenstein, W. R.** Cell junction and cyclic AMP. III. Promotion of junctional membrane permeability and junctional membrane particles in a junction-deficient cell type, *J. Membr. Biol.*, 63, 133, 1981.

4. **Bennett, M. V. L. and Goodenough, D. A.**, Gap junctions, electrotonic coupling and intercellular communication, *Neurosci. Res. Prog. Bull.*, 16, 373, 1978.

5. **Bennett, M. V. L., Verselis, V., White, R. L., and Spray, D. C.**, Gap junction conductance: gating, in *Gap Junctions*, Vol. 7, Hertzberg, E. L. and Johnson, R. G., Eds., Alan R. Liss, New York, 1988.

6. **Blackshaw, S. E. and Warner, A. E.**, Low resistance junctions between mesoderm cells during development of trunk muscles, *J. Physiol. (London)*, 255, 209, 1976.

7. **Burt, J. M. and Spray, D. C.**, Single-channel events and gating behavior of the cardiac gap junction channel, *Proc. Natl. Acad. Sci. U.S.A.*, 85, 3431, 1988.

8. **Chiarugi, V. P., Vanucchi, S., and Urbano, P.**, Exposure of trypsin-removable sulphated polyanions on the surface of normal and virally transformed BHK^{21}/C13 cells, *Biochim. Biophys. Acta*, 345, 283, 1974.

9. **Chow, I. and Cohen, M. W.**, Developmental changes in the distribution of acetylcholine receptors in the myotomes of *Xenopus laevis*, *J. Physiol. (London)*, 339, 553, 1983.

10. **Chow, I. and Poo, M-M.**, Formation of electrical coupling between embryonic *Xenopus* muscle cells in culture, *J. Physiol. (London)*, 346, 181, 1984.

11. **Chow, I. and Young, S. H.**, Opening of single gap junction channels during formation of electrical coupling between embryonic muscle cells, *Devel. Biol.*, 122, 332, 1987.

12. **Culp, L. A. and Black, P. H.**, Release of macromolecules from BALB/c mouse cell lines treated with chelating agents, *Biochemistry*, 11, 2161, 1972.

13. **Délèze, J. and Hervé, J. C.**, Effect of several uncouplers of cell- to-cell communication on gap junction morphology in mammalian heart, *J. Membr. Biol.*, 74, 203, 1983.

14. **De Mello, W. C.**, Effect of intracellular injection of cAMP on the electrical coupling of mammalian cardiac cells, *Biochem. Biophys. Res. Commun.*, 119, 1001, 1984.

15. **De Mello, W. C.**, Interaction of cyclic AMP and Ca in the control of electrical coupling in heart fibers, *Biochim. Biophys. Acta*, 888, 91, 1986.

16. **De Mello, W. C.**, Modulation of junctional permeability, in *Cell-to-Cell Communication*, De Mello, W. C., Ed., Plenum Press, New York, 1987, 29.

17. **De Mello, W. C. and Van Loon, P.**, Influence of cyclic nucleotides on junctional permeability in atrial muscle, *J. Mol. Cell. Cardiol.*, 15, 637, 1987.

18. **Epstein, M. L., Sheridan, J. D., and Johnson, R. G.**, Formation of low resistance junctions in vitro in the absence of protein synthesis and ATP production, *Exp. Cell Res.*, 104, 25, 1977.

19. **Flagg-Newton, J. L., Dahl, G., and Loewenstein, W. R.**, Cell junction and cAMP. I. Upregulation of junctional membrane permeability and junctional membrane particles by administration of cyclic nucleotides or phosphodiesterase inhibitor, *J. Membr. Biol.*, 63, 105, 1981.

20. **Furshpan, E. J. and Potter, D. D.**, Transmission of the giant synapse of the crayfish, *J. Physiol. (London)*, 145, 289, 1959.

21. **Gilula, N. B.**, Cell-to-cell communication and development, in *The Cell Surface: Mediator of Developmental Processes*, Subtelny, S. and Wessells, N.K., Eds., Academic Press, New York, 1980, 23.

22. **Hamill, O. P., Marty, A., Neher, E., Sakman, B., and Sigworth, F. J.**, Improved patch-clamp techniques for high-resolution current recording from cells and cell-free patches, *Pflug. Arch.*, 391, 85, 1981.

23. **Ito, S., Sato, E., and Loewenstein, W. R.**, Studies on the formation of a permeable cell membrane junction. I. Coupling under various conditions of membrane contact. Colchicine, cytochalasin B, dinitrophenol, *J. Membr. Biol.*, 19, 305, 1974.

24. **Johnson, R., Hammer, M., Sheridan, J., and Revel, J.-P.**, Gap junction formation between reaggregated Novikoff hepatoma cells, *Proc. Natl. Acad. Sci. U.S.A.*, 71, 4536, 1974.

25. **Johnston, M. F., Simon, S. A., and Ramon, F.**, Interaction of anaesthetics with electrical synapses, *Nature*, 286, 498, 1980.

26. **Kanno, Y. and Loewenstein, W. R.**, Intercellular diffusion, *Science*, 143, 959, 1964.

27. **Kanno, Y. and Loewenstein, W. R.**, Cell-to-cell passage of large molecules, *Nature*, 212, 629, 1966.

28. **Keane, R. W., Mehta, P. P., Rose, B., Honig, L. S., Loewenstein, W. R., and Rutishauser, U.,** Neural differentiation, NCAM-mediated adhesion, and gap junctional communication in neuroectoderm. A study *in vitro, J. Cell Biol.,* 106, 1307, 1988.

29. **Knudsen, K. A. and Horwitz, A. F.,** Toward a mechanism of myoblast fusion, in *J. Supramolec. Struct. Cell Surf. Carbohydr. Biol. Recog.,* 563, 1978.

30. **Lasater, E. M. and Dowling, J. E.,** Dopamine decreases conductance of the electical junctions between cultured retinal horizontal cells, *Proc. Natl. Acad. Sci. U.S.A.,* 82, 3025, 1985.

31. **Loewenstein, W. R.,** Junctional intercellular communication: the cell-to-cell membrane channel, *Physiol. Rev.,* 61, 829, 1981.

32. **Loewenstein, W. R., Kanno, Y., and Socolar, S. J.,** Quantum jumps of conductance during formation of membrane channels at cell-cell junction, *Nature,* 274, 133, 1978.

33. **McLachlin, J., Caveney, S., and Kidder, G. M.,** Control of gap junction formation in early mouse embryo, *Devel. Biol.,* 98, 155, 1983.

34. **Neyton, J. and Trautmann, A.,** Single-channel currents of an intercellular junction, *Nature,* 317, 331, 1985.

35. **Neyton, J. and Trautmann, A.,** Acetylcholine modulation of the conductance of intercellular junctions between rat lacrimal cells, *J. Physiol. (London),* 377, 283, 1986.

36. **O'Lague, P. and Delan, H.,** Low resistance junctions between normal and between virus transformed fibroblasts in tissue culture, *Exp. Cell Res.,* 86, 374, 1974.

37. **Oliveira-Castro, G. and Loewenstein, W. R.,** Junctional membrane permeability: effect of divalent cations, *J. Membr. Biol.,* 5, 51, 1971.

38. **Olmsted, J. B. and Borisy, G. G.,** Microtubules, *Annu. Rev. Biochem.,* 42, 507, 1973.

39. **Peracchia, C.,** Calcium effects on gap junction structure and coupling, *Nature,* 271, 669, 1978.

40. **Peracchia, C. and Bernardini, G.,** Gap junction structure and cell-to-cell coupling regulation: is there a calmodulin involvement?, *Fed. Proc.,* 43, 2681, 1984.

41. **Pitts, J. D. and Simms, J. W.,** Permeability of junctions between animal cells. Intercellular transfer of nucleotides, but not of macromolecules, *Exp. Cell Res.,* 104, 153, 1977.

42. **Rabito, C. A., Jarrell, J. A., and Scott, J. A.,** Gap junctions and synchronization of polarization process during epithelial reorganization, *Am. J. Physiol.,* 22, C329, 1987.

43. **Rash, J. E. and Fambrough, D.,** Ultrastructural and electrophysiological correlates of cell coupling and cytoplasmic fusion during myogenesis *in vitro, Devel. Biol.,* 30, 166, 1973.

44. **Rink, T. J., Tsien, R. Y., and Warner, A. E.,** Free calcium in *Xenopus* embryos measured with ion-selective microelectrodes, *Nature,* 283, 658, 1980.

45. **Robertson, J. D.,** Ultrastructure of two invertebrate synapses, *Soc. Exp. Biol. Med.,* 82, 219, 1953.

46. **Rose, B. and Loewenstein, W. R.,** Junctional membrane permeability. Depression by substitution of Li for extracellular Na, and by long term lack of Ca and Mg, restoration by cell repolarization, *J. Membr. Biol.,* 5, 20, 1971.

47. **Rose, B. and Loewenstein, W. R.,** Permeability of cell junction depends on local cytoplasmic calcium activity, *Nature,* 254, 250, 1975.

48. **Saez, J. C., Spray, D. C., Nairn, A. C., Hertzberg, E., Greengard, P., and Bennett, M. V. L.,** cAMP increases junctional conductance and stimulates phosphorylation of the 27-kDa principal gap junction polypeptide, *Proc. Natl. Acad. Sci. U.S.A.,* 83, 2473, 1986.

49. **Safranyos, R. G. A. and Caveney, S.,** Rates of diffusion of fluorescent molecules via cell-to-cell membrane channels in developing tissue, *J. Cell Biol.,* 100, 736, 1985.

50. **Schmalbruch, H.,** Skeletal muscle fibers of newborn rats are coupled by gap junctions, *Devel. Biol.,* 91, 485, 1982.

51. **Sheridan, J. D.,** Dye movement and low resistance junctions between reaggregated embryonic cells, *Devel. Biol.,* 26, 627, 1971.

52. **Sheridan, J. D. and Atkinson, M. M.,** Physiological roles of permeable junctions: some possiblities, *Annu. Rev. Physiol.,* 47, 337, 1985.

53. **Socolar, S. J. and Loewenstein, W. R.,** Methods for studying transmission through permeable cell-to-cell junctions, in *Methods in Membrane Biology,* Vol 10., Korn, E. D., Ed., Plenum Press, New York, 1979, 123.

54. **Spray, D. C. and Bennett, M. V. L.,** Physiology and pharmacology of gap junctions, *Rev. Physiol.,* 47, 281, 1985.

55. **Spray, D. C., Harris, A. L., and Bennett, M. V. L.,** Voltage dependence of junctional conductance in early amphibian embryos, *Science,* 204, 432, 1979.

56. **Spray, D. C., Harris, A. L., and Bennett, M. V. L.,** Gap junctional conductance is a simple sensitive function of intracellular pH, *Science,* 211, 712, 1981.

57. **Stewart, W. W.,** Functional connections between cells as revealed by dye coupling with a highly fluorescent naphthalimide tracer, *Cell,* 4, 741, 1978.

58. **Turin, L. and Warner, A. E.,** Carbon dioxide reversibility abolishes ionic communication between cells of early amphibian embryo, *Nature,* 270, 56, 1977.

59. **Turin, L. and Warner, A. E.,** Intracellular pH in Xenopus embryos: its effects on current flow between blastomeres, *J. Physiol. (London),* 300, 489, 1980.

60. **Veenstra, R. D. and DeHaan, R. L.,** Measurement of single channel currents from cardiac gap junctions, *Science,* 233, 972, 1986.

61. **Weber, K. and Osborn, M.,** Microtubule and intermediate filament networks in cells viewed by immunofluorescence microscopy, in *Cytoskeletal Elements and Plasma Membrane Organization,* Poste, G. and Nicolson, G. L., Eds., Elsevier/North-Holland, Amsterdam, 1981, 1.

62. **Weldon, P. R. and Cohen, M. W.,** Development of synaptic ultrastructure at neuromuscular contacts in an amphibian cell culture system, *J. Neurocytol.,* 8, 239, 1979.

63. **Werner, R., Miller, T., Azarnia, R., and Dahl, G.,** Translation and functional expression of cell-cell channel mRNA in *Xenopus* oocytes, *J. Membr. Biol.,* 87, 253, 1985.

64. **Wilson, L., Anderson, K., Grisham, L., and Chin, L.,** Biochemical mechanisms of action of microtubule inhibitors, in *Microtubules and Microtubule Inhibitors,* Borges, M. and De Brabander, M., Eds., Elsevier/North-Holland, Amsterdam, and American Elsevier, New York, 1975, 103.

65. **Young, S. H. and Chow, I.,** Modulation of formation and maintenance of electrical coupling between muscle cells in culture, Int. Conf. Gap Junctions (Abstr.), 37, 1987.

66. **Zampighi, G.,** Gap junction structure, in *Cell-to-cell Communication,* De Mello, W. C., Ed., Plenum Press, New York, 1987, 1.

Chapter 4

SELECTIVE FORMATION AND MODULATION OF ELECTRICAL SYNAPSES

Grant M. Carrow and Irwin B. Levitan

TABLE OF CONTENTS

I. INTRODUCTION

The complex functions of the nervous system depend upon both the specificity and plasticity of the synaptic connections that constitute the communication links in neural circuitry. How synaptic connections are specified during assembly of the nervous system and how they are later modified to bring about learning and memory are among the most fascinating problems in neurobiology. We know that synaptic specificity arises during development as part of a series of circuit assembly processes beginning with neural migration, continuing through axon outgrowth and pathfinding, and culminating in target recognition and synapse formation. Yet, we do not fully understand what constitutes the information used by neurons to achieve the final pattern of connections. Evidently, there exists a map of positional and directional cues used by growing axons, yet we do not know what constitutes the map nor how it is read. Moreover, it is evident that once the circuitry is hardwired, its proper functioning depends upon its capacity to undergo modification, and one way in which this might be accomplished is by modulation of individual synapses. The cellular and molecular mechanisms by which synaptic connections may be modified are just beginning to be understood. Our focus in this chapter is on how neurons may specify their connections during the final stages of neural circuit assembly and how those connections may be modified once established. As a basis for discussion, we use as an example work from our laboratory on a system of cultured neurons in which both problems may be addressed simultaneously. This area of research is in its infancy, having become practical only with the recent advent of culturing techniques that permit access to the site of intercellular communication in the nervous system: the synapse.

Intercellular communication in the nervous system is mediated by the transfer of chemical and electrical signals via two distinct types of synaptic specialization. For chemical synapses, molecules constitute the sole communication link between neurons, providing for the indirect, unidirectional transfer of information. Thus, electrical information from the presynaptic cell is coded by the quantity and pattern of neurotransmitter released while receptors in the postsynaptic cell decode the chemical signals and transduce them back into electrical signals for further transmission. By contrast, electrical synapses provide a communication link for the direct transfer of electrical and chemical signals between neurons. The low resistance pathway between cells, the gap junction, permits electrotonic spread of electrical signals as well as passage of ions and small molecules in either direction. Thus, the neurons are both electrically and metabolically coupled. Nonexcitable cells share this latter type of communication port, as shown elsewhere in this volume, although the duality of chemical and electrical connectivity is found only in excitable cells.

Both types of synapse constitute the unit building block for networks in the nervous system and are the ultimate foci for the study of interneuronal communication. Yet, because of their small size and the complexity of neural tissue, the sites of synaptic connections often are not accessible to experimental manipulation. Cell culture provides a more tractable means for investigating the processes involved in synaptogenesis and synapse modulation. We have focused on electrical synapses because in our hands they form more reliably in culture than do chemical synapses and the efficacy of synaptic connections may be more readily quantified. Our beast of choice is the marine mollusk, *Aplysia californica,* whose central nervous system is well suited to physiological and biochemical analysis. The nervous system of *Aplysia* has thus yielded fundamental insights into the regulation of synapses and the modulation of membrane ion currents and promises to provide basic information on mechanisms of synaptogenesis and the modulation of gap junctions.

II. SPECIFIC NEURONAL INTERACTIONS

A. SELECTIVE AXON FASCICULATION

Specificity during the early stages of neural circuit assembly, including pathfinding by growth cones and axon fasciculation, appears to depend upon guidance cues encoded in cell surface proteins that differentially label axon pathways. The first neuronal growth cones, or pioneers, in the central nervous system follow pathways that appear to be defined by non-neuronal cells whereas later growth cones follow characteristic pathways defined by pioneers.[1-4] The most convincing evidence for the specificity of guidance pathways by pioneer neurons in the central nervous system is the observation that ablation of presumed pioneer neurons leads to failure of growth cones to grow on any inappropriate axons.[4,5] Moreover, specificity is expressed *in vitro* when neurons are challenged to contact appropriate as well as inappropriate targets. For example, neurites of embryonic retinal explants grow freely on one another as do those of sympathetic ganglion explants. However, retinal and sympathetic neurites avoid one another in mixed culture.[6,7] Time-lapse observations reveal that the growth cones from either type of neuron readily cross homotypic neurites (in retinal-retinal or sympathetic-sympathetic combinations) but collapse upon contacting heterotypic neurites (retinal-sympathetic combinations).[8] Also, when retinal neurons are given a choice of axon pathways for fasciculation in a Y-maze in culture, they often show specificity that correlates with that *in vivo*.[9,10] Specificity in pathfinding is also apparent in the periphery, although the role of pioneer neurons in specifying pathways is less clear.[11-14]

Molecules that are candidates for mediating the specificity of axonal interactions have been identified by generating monoclonal antibodies to cell membranes and demonstrating differential expression of the antigens on specific subsets of axons. Thus, some cell surface glycoproteins whose expression is restricted to subsets of axons have been identified,[15-20] although their functions have not been determined. Monoclonal antibodies to oligosaccharides show that some lactoseries and globoseries glycoconjugates are expressed on distinct subsets of dorsal root ganglion neurons in vivo and *in vitro*.[21] Moreover, antibodies to lactose-binding lectins label the same subsets of sensory neurons that express their complementary oligosaccharide ligands, suggesting a role in cell recognition for these molecules.[22,23] Neuron-specific cell adhesion molecules (N-CAM) have also been implicated in axon guidance and synapse formation.[20,24-26] Although these molecules are not differentially expressed on subsets of neurons, a graded distribution or heterogeneities of adhesiveness might provide positional information.

B. SELECTIVE SYNAPTOGENESIS

It is possible that synaptic specificity arises solely from selective fasciculation by the guidance of neurons to the appropriate targets and that further positional or identification cues are unnecessary. Yet, different subsets of similar neurons that follow a common pathway and innervate a common target may still demonstrate specific differences in their synapses. An example of this phenomenon is afforded by the circuitry of the monosynaptic stretch reflex of vertebrates. Stretch-sensitive sensory neurons make stronger synaptic connections with related (homonymous) motoneurons than with unrelated (heteronymous) motoneurons in the same region of the spinal cord.[27-30] In the autonomic nervous system, neurons in a given sympathetic ganglion are innervated most strongly by preganglionic axons arising from a characteristic spinal cord segment.[31] Moreover, a ganglion transplanted to an inappropriate level in the sympathetic chain is reinnervated in a manner that often approximates its original segmental innervation.[32] The specificity of connections in these cases are likely signatures of the specific cellular interactions that occurred during synaptogenesis. Substantial evidence for selective synaptogenesis has also been obtained from experiments *in vitro,* discussed in detail below. The molecules that mediate synaptic selectivity remain to be identified, although some indication that cell surface molecules may be differentially expressed at synapses is given by the observations that lectins specifically label neuromuscular junction, but not extrajunctional regions[33,34] and bind preferentially at the synaptic cleft in synaptosomes.[35-37]

C. SPECIFIC CONNECTIONS IN CULTURE

Studies of the specificity of synaptogenesis have been hampered by the limitations to observing and manipulating the process directly *in vivo.* These constraints have been alleviated by the recent advances in primary culture of neurons from the central nervous system and the finding that neurons form specific synaptic connections *in vitro.* Certainly, the caveat remains that synaptogenesis *in vitro* may not exibit all the features of the process *in vivo.* Nevertheless, the combination of ready access to the sites of synaptic connection and the ability to present neurons with appropriate as well as inappropriate targets have allowed us to address problems that are intractable *in vivo.*

Neurons in culture show specificity in the formation of both chemical and electrical synapses, often mirroring the situation *in vivo.* Nerve-muscle synapses have been reconstituted *in vitro* with dissociated motoneurons and muscle cells from vertebrates[38,39] and *Drosophila.*[40] The best evidence for specific interneuronal synapses is from invertebrates, whose large, identified cells facilitate *in vivo* and *in vitro* comparisons of synaptology. For example, L10 neurons of *Aplysia* make inhibitory cholinergic synapses in culture on their *in vivo* targets, left upper quadrant neurons, but not with inappropriate targets, right upper quadrant neurons, even though both sets of targets have acetylcholine (ACh) receptors.[41] An elementary gill withdrawal circuit when reconstituted with sensory neurons and motoneurons in culture shows short-term homosynaptic depression and heterosynaptic facilitation.[42] In some cases where neurons make specific chemical synapses in culture, some resemble those *in vivo* while others are seen only in culture.[43-49]

Electrical synapses also demonstrate specificity in culture. For example, P sensory neurons from the leech couple with L motoneurons, but not with Retzius cells even though the latter couple with one another.[44] In *Aplysia,* neurons from the bag cell clusters in the abdominal ganglion are joined by gap junctions *in vivo* and readily form electrical connections *in vitro.*[45,50,51] Similarly, neurons from the buccal ganglion are, in many cases, electrically coupled *in vivo*[52] and *in vitro.*[45,53] In mixed cultures, however, buccal and bag neurons form weak or undetectable electrical connections with each other.[45,51,54] Because buccal and bag neurons are fully competent to form strong electrical connections with homotypic partners, the weak or absent heterotypic coupling must reflect selectivity in the formation of electrical synapses during synaptogenesis. This selectivity is manifested as a difference in synaptic efficacy,[54] similar to the situation in the stretch reflex and sympathetic circuits described above.

Similar observations of selectivity in the formation of electrical connections between homotypic and heterotypic partners have been made both *in vivo* and *in vitro*. In the tiger salamander retina, rods are more effectively coupled to adjacent rods than they are to adjacent cones; adjacent cones are not directly coupled to one another.[55,56] These differences appear to be important for the integrative functions of photoreceptors.[57] Examples abound *in vitro* because of the virtually unlimited capacity to pair heterotypic cells. Thus, epithelial cells from rabbit lens and rat liver *in vitro* show lower coupling coefficients in the heterotypic combination than in the homotypic combinations.[58] In other cases of heterotypic pairings in culture no electrical coupling was detected despite the competence of the cells to couple in homotypic combinations.[45,51,59-61] Failure to detect any coupling between competant cells when paired in heterotypic combinations may in part be a function of the culture conditions;[51,54,62-65] nevertheless, there remains a difference in coupling efficacy that reflects the selectivity of formation of the connections. There is at least one exception, however, to the pattern of specificity in electrical connections since mixed cultures of rat ovarian granulosa cells and mouse myocardial cells show a high level of heterotypic coupling.[66]

III. SELECTIVE SYNAPSE FORMATION BETWEEN *APLYSIA* NEURONS

As mentioned above, experiments in this laboratory demonstrated that specificity of electrical synapse formation is expressed by *Aplysia* buccal and bag neurons in mixed culture.[45,51,54] We are interested in using this phenomenon as a simple model with which to address problems regarding target recognition and the regulation of synaptic connections.

In order to better understand the observations with buccal and bag neurons, we investigated the extent to which synaptic specificity applies to neurons derived from other areas of the central nervous system of *Aplysia*. Thus, we placed various combinations of neurons from the several ganglia in primary cell culture in either of two configurations: (1) somata separated by 50 to 300 μm with contacts limited to regenerated neurites (Figure 1A) or (2) somata directly apposed (Figure 1B). Apposition of somata allows for electrical coupling in the absence of neurite outgrowth and increases the frequency with which cell pairs are electrically coupled;[44,54] it does not, however, increase the coupling coefficient as has been reported previously for bag cells in culture.[50] After allowing for neurite outgrowth (if desired) and synapse formation (usually 2 to 3 days) each cell of a pair is penetrated with microelectrodes and either current clamped to measure coupling coefficients or voltage clamped to measure membrane and junctional conductances.[54,67,68] We use a single microelectrode in each cell for clamping by rapidly switching between current injection and voltage monitoring (Figure 2A).[69] This facilitates long-term recording since it reduces the chances of losing a recording when compared to two-electrode clamp configurations and does not involve the intracellular perfusion that accompanies whole cell recording. With a dual

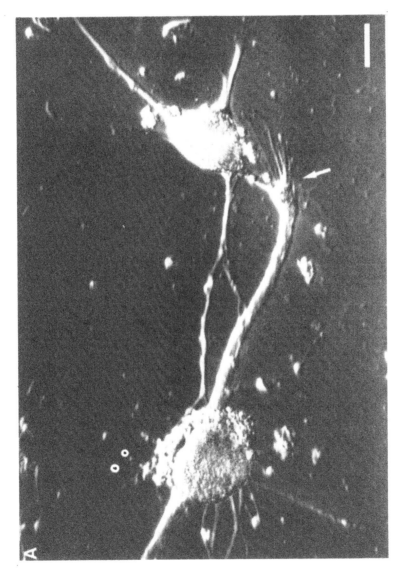

FIGURE 1. Configurations of *Aplysia* neurons in primary cell culture. Light photomicrographs of neurons with regenerated neurites viewed with Hoffman modulation contrast optics. (A) A separated pair of left upper quadrant neurons from the abdominal ganglion (homoganglionic pair) with neurites in contact; a growth cone with long filopodia is indicated by the arrow. (B) An apposed pair consisting of a buccal and a pleural neuron (heteroganglionic pair). Calibration bars equal 50 μm.

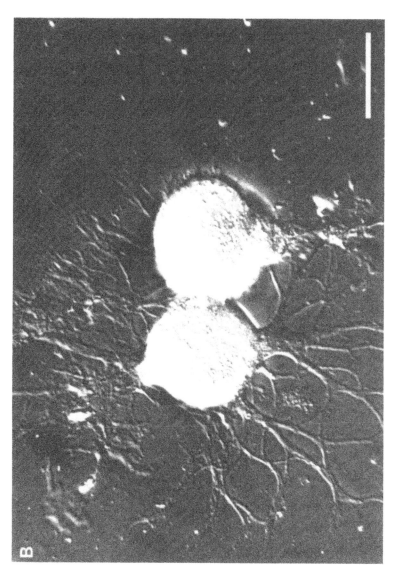

FIGURE 1 (continued)

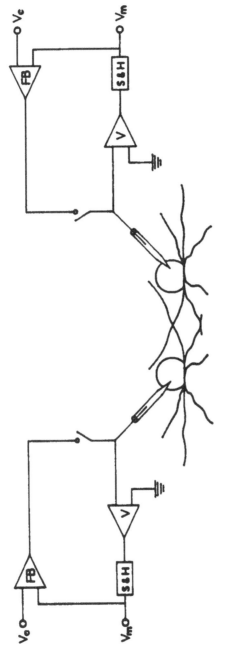

FIGURE 2.   Current clamp and voltage clamp recording from pairs of neurons in culture. (A) Configuration for single-electrode current and voltage clamping. The electrode is continuously switched between current injection and voltage monitoring. FB, feedback amplifier; S & H, sample-and-hold device; V, voltage follower; V_c, command voltage; V_m, voltage monitor. (B) Measurement of junctional current (I_2) in a postsynaptic (follower) buccal neuron in response to a command voltage step (V_1) in a presynaptic buccal neuron. Note that because the voltage of the postsynaptic cell (V_2) is clamped at a constant voltage during the pulse, the only current in that cell is due to the macroscopic junctional conductance (G_j). Nonjunctional currents (I_1) are apparent in the presynaptic cell. (C) Current-voltage (I-V) curves for the connection in B. G_j (slope) is 63 nS; G_j is (1) linear over a wide range of voltages and is thus not voltage dependent; (2) not dependent upon the sign of the current injected and is thus nonrectifying; and (3) not dependent upon the holding potential of the cell (abcissa intersect for each of the curves: −50 (●), −70 (▲), and −90 mV (■). The lines represent linear regressions fit to the data.

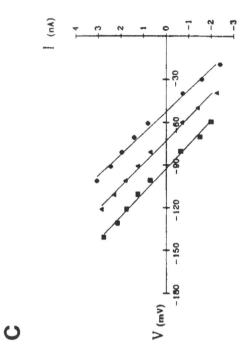

C

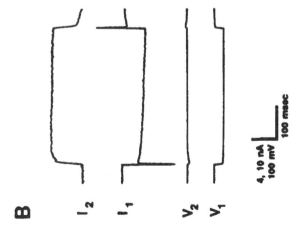

B

FIGURE 2 (continued)

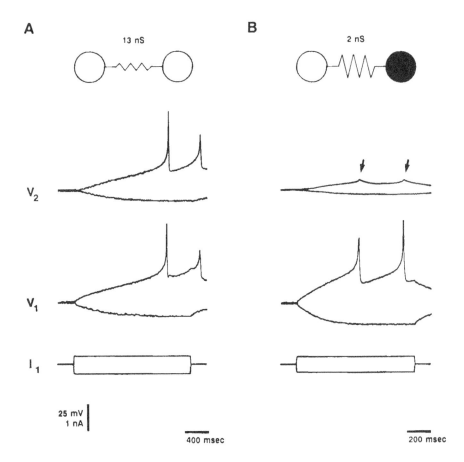

FIGURE 3. Comparison of electrical coupling at homoganglionic and heteroganglionic synapses. The cells were current clamped at –60 to –70 mV; one cell of a pair (V_1; presynaptic cell) was injected with depolarizing or hyperpolarizing current (I_1) while the change in membrane potential of the other, postsynaptic, cell (V_2) was monitored. (A) A pair of left upper quadrant neurons (homoganglionic) from the abdominal ganglion separated by 100 μm and showing a coupling coefficient (k) of 0.75 and a G_j of 13 nS. Note that the electrotonic postsynaptic potentials (psp's) elicited action potentials in the postsynaptic cell. (B) An apposed buccal-pleural pair (heteroganglionic) with k = 0.25 and G_j = 2 nS (the lower conductance is indicated at top by the larger resistor since the junctional resistance is the inverse of G_j). Here, the psp's (arrows) were not sufficient to evoke action potentials in the postsynaptic cell. (C) I-V curves for a pair of buccal neurons (□; BC/BC; homoganglionic; G_j = 21 nS) and a cerebral-pleural pair (○; CR/PL; heteroganglionic; G_j = 5 nS) voltage clamped as described in Figure 2 and showing lack of voltage dependence and rectification in both types of synapse. (Modified from Carrow, G. M. and Levitan, I. B., *J. Neurosci.*, 9, 1989, in press.)

voltage clamp arrangement the junctional current between two cells may be measured directly and characterized (Figures 2B and C).

When cells are paired in the manner described, those pairs consisting of cells from the same ganglion or cluster ("homoganglionic" pairs) are coupled with high coupling coefficients ranging up to 0.95. By contrast, those pairs consisting of cells from different ganglia in mixed culture ("heteroganglionic" pairs) show low coupling coefficients less than 0.25 (Figure 3).[54] The difference in levels of coupling results from

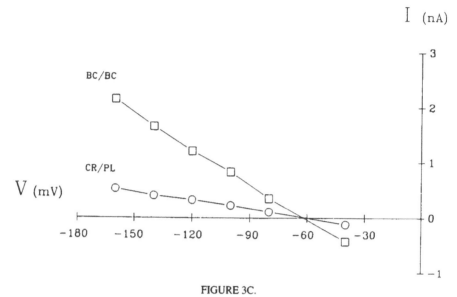

FIGURE 3C.

a higher macroscopic junctional conductance in homoganglionic pairs when compared to heteroganglionic pairs, rather than differences in membrane input impedence. Thus, homoganglionic pairs have macroscopic junctional conductances usually in the range of 10 to 25 nS, and occasionally higher, whereas heteroganglionic pairs have junctional conductances less than 5 nS (Figure 3).[54] The functional significance of these different levels is indicated by the nonlinear dependence of the coupling coefficient on the junctional and nonjunctional conductances of the membrane;[67,70] i.e., for the range of input impedences of these neurons (typically 25 to 100 MΩ), the coupling coefficient is very sensitive to changes in junctional conductance when the latter is low (less than 10 to 20 nS), but is much less sensitive at higher levels of junctional conductance (Figure 4). Thus, junctional conductances of 10 to 25 nS fall right in the range of transition between these two levels of sensitivity and may give rise to coupling coefficients of 0.5 or greater. The functional outcome of the different magnitudes of coupling is that action potentials in a presynaptic (current-injected) cell often evoke action potentials in the postsynaptic (follower) cell of a homoganglionic pair (Figure 3A) but not in the postsynaptic cell of a heteroganglionic pair (Figure 3B). How the differences in magnitude of electrical coupling between the neurons might be related to differences in metabolic coupling is unknown; however, permeability is a linear function of junctional conductance and therefore there may be significant differences in metabolic coupling as well.[59,66,71,72]

The observed dichotomy in synaptic efficacy shows that these neurons have the capacity to distinguish homoganglionic from heteroganglionic partners in culture, suggesting interactions mediated by ganglion-specific cell recognition molecules. It is possible that the recognition system used by these neurons in specific synaptogenesis is the same as that involved in specific axon fasciculation. However, this is unlikely since neurons in heteroganglionic combinations, as in homoganglionic combinations, readily adhere to one another and their axons contact one another. In any case, we now

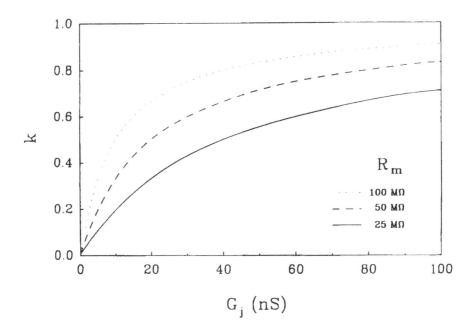

FIGURE 4. Coupling coefficient (k) as a function of junctional conductance (G_j) and membrane input impedence (R_m). Theoretical curves for k vs. G_j are plotted for three values of R_m within the range common for *Aplysia* neurons in culture. The curves were derived from the equation $k = G_j \cdot R_m/(1+G_j \cdot R_m)$ for the simplest case of an isolated pair of electrically coupled cells with equal R_m.[67,70]

have an assay system with which molecules involved in cell recognition may be identified. The relationship of the specificity expressed by these regenerating neurons in culture to the processes active during development *in vivo* remains to be determined. However, the observations of coincidence between communication compartments and developmental compartments in other systems[73] suggests that the differences in electrical coupling observed between *Aplysia* neurons might have a role during gangliogenesis.

IV. MODULATION OF *APLYSIA* ELECTRICAL SYNAPSES

As mentioned above, buccal and bag cells of *Aplysia,* like other heteroganglionic combinations, normally show weak or undetectable coupling.[45,51,54] However, when buccal-bag pairs are grown in the continued presence of the lectin, Concanavalin A (Con A), the frequency of detectably coupled pairs is increased in a dose-dependent manner.[51] This suggests that there is a degree of plasticity inherent in the electrical connections that is revealed by this probe.

Con A is best known for its action as a mitogen for lymphocytes,[74,75] but, along with other lectins, has a broad range of effects on neurons. Other actions of Con A in *Aplysia* include unmasking of a response to L-glutamate,[76] enhancing neurite outgrowth,[51] and increasing the activity of a potassium channel.[77] Con A and other lectins also affect the growth, attachment, or aggregation of other types of neurons,[78-82] and Con A induces neural tissue and cartilage in amphibian embryos.[83,84] Finally, Con A increases depolarization-dependent flux of calcium in PC12 cells[85] and prevents rapid diffusion of ACh receptors and nerve-induced receptor accumulation in amphibian nerve-muscle cultures.[86] The significance of these myriad effects of Con A is unknown, although in at least some of these cases specific receptors are intimated.

As a first step toward determining how Con A alters the electrical coupling between neurons, we allowed heteroganglionic pairs to establish electrical synapses in the absence of either serum or Con A, and then subjected them to acute treatment with Con A while the cells were voltage clamped. We found that the junctional conductance of Con A-treated heteroganglionic pairs increases gradually over several hours until it reaches the level characteristic of homoganglionic pairs (Figure 5A). The already high junctional conductance of homoganglionic pairs is not further increased with Con A treatment (Figure 5A). Thus, it appears that Con A modulates the junctional conductance between these neurons. Figure 5B is a diagramatic summary of the observed synaptic specificity and modulation of junctional conductance among *Aplysia* neurons. Other modulators that increase electrical or dye coupling between cells in some systems include hormones,[64,87,88] a catecholamine,[87] and cAMP and cAMP-dependent protein kinase.[64,89-92] Catecholamines and cAMP decrease coupling in other systems[93-97] as does the *src* gene.[71,98,99] Several of these and other modulators of coupling are discussed elsewhere in this volume.

The mechanism by which Con A modulates the junctional conductance between *Aplysia* neurons remains to be determined. Con A, like other lectins, is a carbohydrate binding protein that can cross-link receptors on cell membranes and can cause cell-cell agglutination.[100,101] However, these properties are apparently not involved in at least some of the biological effects of Con A.[75,76,102] Indeed, the concentrations of Con A effective in modulating junctional conductance between *Aplysia* neurons are nonagglutinating for at least some cells.[103] Moreover, the succinylated dimer of Con A does not cross-link receptors or cause agglutination,[102] yet has activity equivalent to native Con A in increasing junctional conductance.[104] Furthermore, modulation by Con A is blocked by the high concentrations of methyl α-D-mannopyranoside[104] that block Con A binding.[105] Thus, it is likely that the action of Con A is specific, possibly acting via a receptor. In light of the observations suggesting that lectins and lectin receptors may be involved in specific axonal interactions,[22,23] and that synaptic efficacy is regulated by both Con A and endogenous cell recognition cues, it is possible that the lectin receptors suggested by our findings on modulation are, or are closely associated with, the cell recognition molecules that mediate the observed synaptic specificity.

The long time course of action of Con A in increasing junctional conductance suggests that protein synthesis might be involved, as for some other cases of epigenetically induced increases in coupling.[89,106] When heteroganglionic pairs are treated with Con A in the presence of the protein synthesis inhibitor, anisomycin, no increase in junctional conductance is observed (Figure 6).[107] However, the effect is reversible, since removal of anisomycin and continued exposure of the pairs to Con A results in a subsequent increase in junctional conductance. Thus it appears that at least part of the modulation of junctional conductance by Con A is dependent upon protein synthesis. It remains to be determined whether the relevant proteins are part of the gap junction channel itself or are regulators of the channel.

A

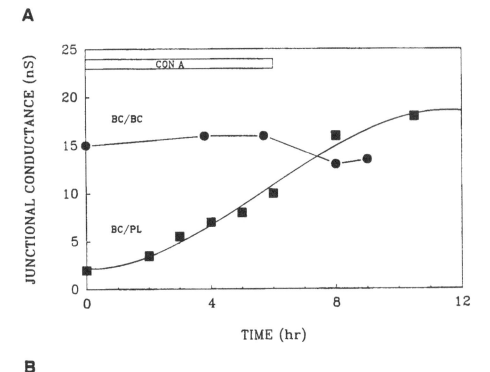

TIME (hr)

B

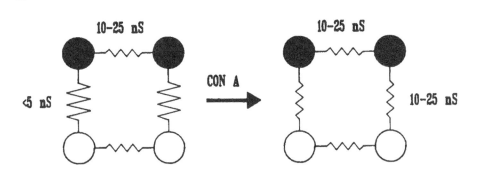

FIGURE 5. Modulation of junctional conductance by Con A. (A) Junctional conductance for a buccal-pleural pair (■; BC/PL; heteroganglionic) as a function of time after starting bath application of Con A (at $t = 0$). G_j increased an order of magnitude 10 h after initiating perfusion with 100 nM Con A for the time indicated by the horizontal bar. The effect was not reversible with 4.5 h wash. The line represents a third order polynomial fitted to the data. For comparison, the time course of G_j measurements from a buccal-buccal pair (●; BC/BC; homoganglionic) treated similarly is plotted. There was no change in G_j. (Modified from Carrow, G. M. and Levitan, I. B., *J. Neurosci.*, 9, 1989, in press.) (B) Diagrammatic representation of the specificity of electrical coupling between cultured neurons and its modulation by Con A. Untreated homoganglionic pairs (indicated by equivalent shading) have high G_j (small resistors) whereas heteroganglionic pairs have low G_j (large resistors). After treatment with Con A, all connections are characterized by high G_j (small resistors). Two of the four possible heteroganglionic connections in a tetrad are omitted for clarity.

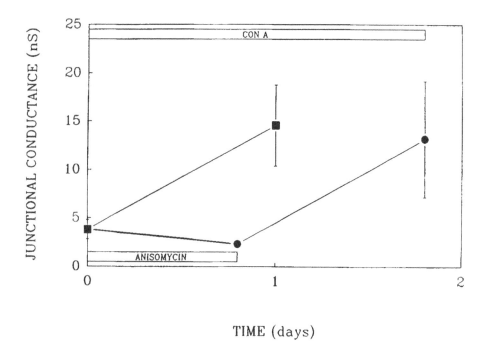

FIGURE 6. Protein synthesis dependence of the Con A-induced increase in junctional conductance. As shown in Figure 5, heteroganglionic pairs of neurons treated with 100 nM Con A (■) increase G_j in less than 1 day (n ≥ 20; data from Carrow, G. M. and Levitan, I. B., *J. Neurosci.,* 9, 1989, in press.) However, heteroganglionic pairs treated with 500 nM Con A in the presence of 10 μM anisomycin (●), a protein synthesis inhibitor, for a similar time period do not increase G_j (n = 12); the apparent decrease in G_j is not significant (p >> 0.05). After removal of anisomycin and continued exposure to Con A, the junctional conductance increases, indicating that the block is reversible (n = 4). Plotted are the means and SE; for one point (●) the error bars are shorter than the symbol radius.

V. PROSPECTUS

In a simple system, we are able to predict the patterns of cell-cell connectivity, quantify the efficacy of synaptic connections, and alter synaptic efficacy with a specific ligand. It may thus be feasible to examine simultaneously the major events during and after synaptogenesis, namely cell-cell recognition, synapse formation, and synaptic modulation. Con A should prove to be a valuable tool for uncovering the mechanisms involved in intercellular communication during these crucial events in neural development.

ACKNOWLEDGMENTS

The research described here was supported by National Research Service Award fellowship F32-HD06739 to G.M.C. and National Institutes of Health grant NS25366 to I.B.L.

REFERENCES

1. **Singer, M., Nordlander, R. H., and Egar, M.,** Axonal guidance during embryogenesis and regeneration in the spinal cord of the newt: the blueprint hypothesis of neuronal pathway patterning, *J. Comp. Neurol.,* 185, 1, 1979.

2. **Goodman, C. S., Raper, J. A., Ho, R. K., and Chang, S.,** Pathfinding by neuronal growth cones in grasshopper embryos, in *N.Y. Symp. Soc. Dev. Biol.,* Vol. 40: *Developmental Order: Its Origin and Regulation,* Subtelny, S. and Green, P. B., Eds., Alan R. Liss, 1982, 275.

3. **Raper, J. A., Bastiani, M., and Goodman, C. S.,** Pathfinding by neuronal growth cones in grasshopper embryos. II. Selective fasciculation onto specific axonal pathways, *J. Neurosci.,* 3, 31, 1983.

4. **Kuwada, J. Y.,** Cell recognition by neuronal growth cones in a simple vertebrate embryo, *Science,* 233, 740, 1986.

5. **Raper, J. A., Bastiani, M. J., and Goodman, C. S.,** Pathfinding by neuronal growth cones in grasshopper embryos. IV. The effects of ablating the A and P axons upon the behavior of the G growth cone, *J. Neurosci.,* 4, 2329, 1984.

6. **Bray, D., Wood, P., and Bunge, R. P.,** Selective fasciculation of nerve fibres in culture, *Exp. Cell Res.,* 130, 241, 1980.

7. **Kapfhammer, J. P., Grunewald, B. E., and Raper, J. A.,** The selective inhibition of growth cone extension by specific neurites in culture, *J. Neurosci.,* 6, 2527, 1986.

8. **Kapfhammer, J. P. and Raper, J. A.,** Interactions between growth cones and neurites growing from different neural tissues in culture, *J. Neurosci.,* 7, 1595, 1987.

9. **Bonhoeffer, F. and Huf, J.,** Position-dependent properties of retinal axons and their growth cones, *Nature,* 315, 409, 1985.

10. **Walter, J., Henke-Fahle, S., and Bonhoeffer, F.,** Avoidance of posterior tectal membranes by temporal retinal axons, *Development,* 101, 909, 1987.

11. **Lance-Jones, C. and Landmesser, L.,** Pathway selection by embryonic chick motoneurons in an experimentally altered environment, *Proc. R. Soc. London Ser. B,* 214, 19, 1981.

12. **Bentley, D. and Keshishian, H.,** Pathfinding by peripheral pioneer neurons in grasshoppers, *Science,* 218, 1082, 1982.

13. **Tosney, K. W. and Landmesser, L. T.,** Specificity of early motoneuron growth cone outgrowth in the chick embryo, *J. Neurosci.,* 5, 2336, 1985.

14. **Eisen, J. S., Myers, P. Z., and Westerfield, M.,** Pathway selection by growth cones of identified motoneurones in live zebra fish embryos, *Nature,* 320, 269, 1986.

15. **McKay, R. D. G., Hockfield, S., Johansen, J., Thompson, I., and Frederiksen, K.,** Surface molecules identify groups of growing axons, *Science,* 222, 788, 1983.

16. **Yamamoto, M., Boyer, A. M., Crandall, J. E., Edwards, M., and Tanaka, H.,** Distribution of stage-specific neurite-associated proteins in the developing murine nervous system recognized by a monoclonal antibody, *J. Neurosci.,* 6, 3576, 1986.

17. **Bastiani, M. J., Harrelson, A. L., Snow, P. M., and Goodman, C. S.,** Expression of fasciclin I and II glycoproteins on subsets of axon pathways during neuronal development in the grasshopper, *Cell,* 48, 745, 1987.

18. **Patel, J., Snow, P. M., and Goodman, C. S.,** Characterization and cloning of fasciclin III: a glycoprotein expressed on a subset of neurons and axon pathways in *Drosophila, Cell,* 48, 975, 1987.

19. **Dodd, J., Morton, S. B., Karagogeos, D., Yamamoto, M., and Jessell, T. M.,** Spatial regulation of axonal glycoprotein expression on subsets of embryonic spinal neurons, *Neuron,* 1, 105, 1988.

20. **Jessell, T. M.,** Adhesion molecules and the hierarchy of neural development, *Neuron,* 1, 3, 1988.

21. **Dodd, J. and Jessell, T. M.,** Lactoseries carbohydrates specify subsets of dorsal root ganglioin neurons projecting to the superficial dorsal horn of rat spinal cord, *J. Neurosci.,* 5, 3278, 1985.

22. **Dodd, J. and Jessell, T. M.,** Cell surface glycoconjugates and carbohydrate-binding proteins: Possible recognition signals in sensory neurone development, *J. Exp. Biol.,* 124, 225, 1986.

23. **Regan, L. J., Dodd, J., Barondes, S. H., and Jessell, T. M.,** Selective expression of endogenous lactose-binding lectins and lactoseries glycoconjugates in subsets of rat sensory neurons, *Proc. Natl. Acad. Sci. U.S.A.,* 83, 2248, 1986.

24. **Edelman, G. M.,** Cell adhesion molecules in neural histogenesis, *Annu. Rev. Physiol.,* 48, 417, 1986.

25. **Rutishauser, U., Acheson, A., Hall, A. K., Mann, D. M., and Sunshine, J.,** The neural cell adhesion molecule (NCAM) as a regulator of cell-cell interactions, *Science*, 240, 53, 1988.

26. **Jacobson, M.,** Neural cell adhesion molecule (NCAM) expression in *Xenopus* embryos during formation of central and peripheral neural maps, in *The Making of the Nervous System*, Parnavelas, J. G., Stern, C. D., and Stirling, R. V., Eds., Oxford University Press, Cambridge, 1988, 128.

27. **Scott, J. G. and Mendell, L. M.,** Individual EPSPs produced by single triceps surae Ia afferent fibers in homonymous and heteronymous motoneurons, *J. Neurophysiol.*, 39, 679, 1976.

28. **Nelson, S. G. and Mendell, L. M.,** Projection of single knee flexor Ia fibers to homonymous and heteronymous motoneurons, *J. Neurophysiol.*, 41, 778, 1978.

29. **Lichtman, J. W. and Frank, E.,** Physiological evidence for specificity of synaptic connections between individual sensory and motor neurons in the brachial spinal cord of the bullfrog, *J. Neurosci.*, 4, 1745, 1984.

30. **Lichtman, J. W., Jhaveri, S., and Frank, E.,** Anatomical basis of specific connections between sensory axons and motor neurons in the brachial spinal cord of the bullfrog, *J. Neurosci.*, 4, 1754, 1984.

31. **Nja, A. and Purves, D.,** Specific innervation of guinea-pig superior cervical ganglion cells by preganglionic fibres arising from different levels of the spinal cord, *J. Physiol.*, 264, 565, 1977.

32. **Purves, D., Thompson, W., and Yip, J. W.,** Re-innervation of ganglia transplanted to the neck from different levels of the guinea-pig sympathetic chain, *J. Physiol.*, 313, 49, 1981.

33. **Sanes, J. R. and Cheney, J. M.,** Lectin binding reveals a synapse-specific carbohydrate in skeletal muscle, *Nature*, 300, 646, 1982.

34. **Scott, L. J. C., Bacou, F., and Sanes, J. R.,** A synapse-specific carbohydrate at the neuromuscular junction: Association with both acetylcholinesterase and a glycolipid, *J. Neurosci.*, 8, 932, 1988.

35. **Bittiger, H. and Schnebli, H. P.,** Binding of concanavalin A and ricin to synaptic junctions of rat brain, *Nature*, 249, 370, 1974.

36. **Cotman, C. W. and Taylor, D.,** Localization and characterization of concanavalin A receptors in the synaptic cleft, *J. Cell Biol.*, 62, 236, 1974.

37. **Kelly, P., Cotman, C. W., Gentry, C., and Nicolson, G. L.,** Distribution and mobility of lectin receptors on synaptic membranes of identified neurons in the central nervous system, *J. Cell Biol.*, 71, 487, 1976.

38. **Fischbach, G. D.,** Synapse formation between dissociated nerve and muscle cells in low density cell culture, *Devel. Biol.*, 28, 407, 1972.

39. **Kidokoro, Y. and Yeh, E.,** Initial synaptic transmission at the growth cone in *Xenopus* nerve-muscle cultures, *Proc. Natl. Acad. Sci. U.S.A.*, 79, 6727, 1982.

40. **Seecof, R. L., Teplitz, R. L., Gerson, I., Ikeda, K., and Donady, J. J.,** Differentiation of neuromuscular junctions in cultures of embryonic *Drosophila* cells, *Proc. Natl. Acad. Sci. U.S.A.*, 69, 566, 1972.

41. **Camardo, J., Proshansky, E., and Schacher, S.,** Identified *Aplysia* neurons form specific chemical synapses in culture, *J. Neurosci.*, 3, 2614, 1983.

42. **Rayport, S. G. and Schacher, S.,** Synaptic plasticity in vitro: cell culture of identified *Aplysia* neurons mediating short-term habituation and sensitization, *J. Neurosci.*, 6, 759, 1986.

43. **O'Lague, P. H., Obata, K., Claude, P., Furshpan, E. J., and Potter, D. D.,** Evidence for cholinergic synapses between dissociated rat sympathetic neurons in cell culture, *Proc. Natl. Acad. Sci. U.S.A.*, 71, 3602, 1974.

44. **Fuchs, P. A., Nicholls, J. G., and Ready, D. F.,** Membrane properties and selective connexions of identified leech neurones in culture, *J. Physiol.*, 316, 203, 1981.

45. **Bodmer, R., Dagan, D., and Levitan, I. B.,** Chemical and electrotonic connections between *Aplysia* neurons in primary culture, *J. Neurosci.*, 4, 228, 1984.

46. **Cooper, E.,** Synapse formation among developing sensory neurones from rat nodose ganglia grown in tissue culture, *J. Physiol.*, 351, 263, 1984.

47. **Schacher, S., Rayport, S. G., and Ambron, R. T.,** Giant *Aplysia* neuron R2 reliably forms strong chemical connections *in vitro*, *J. Neurosci.*, 5, 2851, 1985.

48. **Arechiga, H., Chiquet, M., Kuffler, D. P., and Nicholls, J. G.,** Formation of specific connections in culture by identified leech neurones containing serotonin, acetylcholine and peptide transmitters, *J. Exp. Biol.*, 126, 15, 1986.

49. **Haydon, P. G.,** The formation of chemical synapses between cell-cultured neuronal somata, *J. Neurosci.*, 8, 1032, 1988.

50. **Kaczmarek, L. K., Finbow, M., Revel, J. P., and Strumwasser, F.,** The morphology and coupling of *Aplysia* bag cells within the abdominal ganglion and in cell culture, *J. Neurobiol.,* 10, 535, 1979.

51. **Lin, S. and Levitan, I. B.,** Concanavalin A alters synaptic specificity between cultured *Aplysia* neurons, *Science,* 237, 648, 1987.

52. **Gardner, D.,** Bilateral symmetry and interneuronal organization in the buccal ganglia of *Aplysia, Science,* 173, 550, 1971.

53. **Dagan, D. and Levitan, I. B.,** Isolated identified *Aplysia* neurons in cell culture, *J. Neurosci.,* 7, 736, 1981.

54. **Carrow, G. M. and Levitan, I. B.,** Selective formation and modulation of electrical synapses between cultured *Aplysia* neurons, *J. Neurosci.,* 9, 1989, in press.

55. **Attwell, D. and Wilson, M.,** Behaviour of the rod network in the tiger salamander retina mediated by membrane properties of individual rods, *J. Physiol.,* 309, 287, 1980.

56. **Attwell, D., Wilson, M., and Wu, S. M.,** A quantitative analysis of interactions between photoreceptors in the salamader (*Ambystoma*) retina, *J. Physiol.,* 352, 703, 1984.

57. **Attwell, D., Borges, S., Wu, S. M., and Wilson, M.,** Signal clipping by the rod output synapse, *Nature,* 328, 522, 1987.

58. **Michalke, W. and Loewenstein, W. R.,** Communication between cells of different type, *Nature,* 232, 121, 1971.

59. **Gilula, N. B., Reeves, O. R., and Steinbach, A.,** Metabolic coupling, ionic coupling and cell contacts, *Nature,* 235, 262, 1972.

60. **Fentiman, I., Taylor-Papadimitriou, J., and Stoker, M.,** Selective contact-dependent cell communication, *Nature,* 264, 760, 1976.

61. **Pitts, J. D. and Burk, R. R.,** Specificity of junctional communication between animal cells, *Nature,* 264, 762, 1976.

62. **Flagg-Newton, J. L. and Loewenstein, W. R.,** Cell junction and cyclic AMP. II. Modulations of junctional membrane permeability dependent on serum and cell density, *J. Membr. Biol.,* 63, 123, 1981.

63. **Higgins, D. and Burton, H.,** Electrotonic synapses are formed by fetal rat sympathetic neurons maintained in a chemically-defined culture medium, *J. Neurosci.,* 7, 2241, 1982.

64. **Kessler, J. A., Spray, D. C., Saez, J. C., and Bennett, M. V. L.,** Determination of synaptic phenotype: insulin and cAMP independently initiate development of electrotonic coupling between cultured sympathetic neurons, *Proc. Natl. Acad. Sci. U.S.A.,* 81, 6235, 1984.

65. **Hadley, R. D., Bodnar, D. A., and Kater, S. B.,** Formation of electrical synapses between isolated, cultured *Helisoma* neurons requires mutual neurite elongation, *J. Neurosci.,* 5, 3145, 1985.

66. **Lawrence, T. S., Beers, W. H., and Gilula, N. B.,** Transmission of hormonal stimulation by cell-to-cell communication, *Nature,* 272, 501, 1978.

67. **Bennett, M. V. L.,** Physiology of electrotonic junctions, *Ann. N.Y. Acad. Sci.,* 137, 509, 1966.

68. **Bodmer, R., Verselis, V., Levitan, I. B., and Spray, D. C.,** Electrotonic synapses between *Aplysia* neurons in situ and in culture: aspects of regulation and measurements of permeability, *J. Neurosci.,* 8, 1656, 1988.

69. **Wilson, W. A. and Goldner, M. M.,** Voltage clamping with a single microelectrode, *J. Neurobiol.,* 6, 411, 1975.

70. **Socolar, S. J.,** Appendix. The coupling coefficient as an index of junctional conductance, *J. Membr. Biol.,* 34, 29, 1977.

71. **Sheridan, J. D. and Atkinson, M. M.,** Physiological roles of permeable junctions: some possibilities, *Annu. Rev. Physiol.,* 47, 337, 1985.

72. **Verselis, V., White, R. L., Spray, D. C., and Bennett, M. V. L.,** Gap junctional conductance and permeability are linearly related, *Science,* 234, 461, 1986.

73. **Caveney, S.,** The role of gap junctions in development, *Annu. Rev. Physiol.,* 47, 319, 1985.

74. **Wang, J. L., McClain, D. A., and Edelman, G. M.,** Modulation of lymphocyte mitogenesis, *Proc. Natl. Acad. Sci. U.S.A.,* 72, 1917, 1975.

75. **Wang, J. L. and Edelman, G. M.,** Binding and functional properties of concanavalin A and its derivatives, *J. Biol. Chem.,* 253, 3000, 1978.

76. **Kehoe, J. S.,** Transformation by concanavalin A of the response of molluscan neurones to L-glutamate, *Nature,* 274, 866, 1978.

77. **Lin, S. S., Dagan, D., and Levitan, I. B.,** Concanavalin A modulates a potassium channel in cultured *Aplysia* neurons, *Neuron,* 3, 95, 1989.

78. **DeGeorge, J. J., Slepecky, N., and Carbonetto, S.,** Concanavalin A stimulates neuron-substratum adhesion and nerve fiber outgrowth in culture, *Devel. Biol.,* 111, 335, 1985.

79. **Hashimoto, S., Ikeno, T., and Kuzuya, H.,** Wheat germ agglutinin inhibits the effects of nerve growth factor on the phosphorylation of proteins in PC12h cells, *J. Neurochem.,* 45, 906, 1985.

80. **Chiquet, M. and Acklin, S. E.,** Attachment to Con A or extracellular matrix initiates rapid sprouting by cultured leech neurons, *Proc. Natl. Acad. Sci. U.S.A.,* 83, 6188, 1986.

81. **DeGeorge, J. J. and Carbonetto, S.,** Specificity and valency of Concanavalin A in neuron-substratum adhesion and redistribution of cell surface receptors, *Devel. Biol.,* 119, 45, 1987.

82. **Joubert, R., Caron, M., and Bladier, D.,** Brain lectin-mediated agglutinability of dissociated cells from embryonic and postnatal mouse brain, *Devel. Brain Res.,* 36, 146, 1987.

83. **Takata, K., Yamamoto, K. Y., and Ozawa, R.,** Use of lectins as probes for analyzing embryonic induction, *Wilhelm Roux's Archives,* 190, 92, 1981.

84. **Mikhailov, A. T. and Gogolyuk, N. A.,** Concanavalin A induces neural tissue and cartilage in amphibian early gastrula ectoderm, *Cell Differ.,* 22, 145, 1987.

85. **Greenberg, D. A., Carpenter, C. L., and Messing, R. O.,** Lectin-induced enhancement of voltage-dependent calcium flux and calcium channel antagonist binding, *J. Neurochem.,* 48, 888, 1987.

86. **Kidokoro, Y., Brass, B., and Kuromi, H.,** Concanavalin A prevents acetylcholine receptor redistribution in *Xenopus* nerve-muscle cultures, *J. Neurosci.,* 6, 1941, 1986.

87. **Radu, A., Dahl, G., and Loewenstein, W. R.,** Hormonal regulation of cell junction permeability: upregulation by catecholamine and prostaglandin E, *J. Membr. Biol.,* 70, 239, 1982.

88. **Wolinsky, E. J., Patterson, P. H., and Willard, A. L.,** Insulin promotes electrical coupling between cultured sympathetic neurons, *J. Neurosci.,* 5, 1675, 1985.

89. **Flagg-Newton, J. L., Dahl, G., and Loewenstein, W. R.,** Cell junction and cyclic AMP. I. Upregulation of junctional membrane permeability and junctional membrane particles by administration of cyclic nucleotide or phosphodiesterase inhibitor, *J. Membr. Biol.,* 63, 105, 1981.

90. **Wiener, E. C. and Loewenstein, W. R.,** Correction of cell-cell communication defect by introduction of a protein kinase into mutant cells, *Nature,* 305, 433, 1983.

91. **DeMello, W. C.,** Effect of intracellular injection of cAMP on the electrical coupling of mammalian cardiac cells, *Biochem. Biophys. Res. Commun.,* 119, 1001, 1984.

92. **Saez, J. C., Spray, D. C., Nairn, A. C., Hertzberg, E., Greengard, P., and Bennett, M. V. L.,** cAMP increases junctional conductance and stimulates phosphorylation of the 27-kDa principal gap junction polypeptide, *Proc. Natl. Acad. Sci. U.S.A.,* 83, 2473, 1986.

93. **Teranishi, T., Negishi, K., and Kato, S.,** Dopamine modulates S-potential amplitude and dye-coupling between external horizontal cells in carp retina, *Nature,* 301, 243, 1983.

94. **Piccolino, M., Neyton, J., and Gerschenfeld, H. M.,** Decrease of gap junction permeability induced by dopamine and cyclic adenosine 3′:5′-monophosphate in horizonatal cells of turtle retina, *J. Neurosci.,* 4, 2477, 1984.

95. **Lasater, E. M. and Dowling, J. E.,** Dopamine decreases conductance of the electrical junctions between cultured retinal horizontal cells, *Proc. Natl. Acad. Sci. U.S.A.,* 82, 3025, 1985.

96. **Piccolino, M., Witkovsky, P., and Trimarchi, C.,** Dopaminergic mechanisms underlying the reduction of electrical coupling between horizontal cells of the turtle retina induced by d-amphetamine, bicuculline, and veratridine, *J. Neurosci.,* 7, 2273, 1987.

97. **Lasater, E. M.,** Retinal horizontal cell gap junctional conductance is modulated by dopamine through a cyclic AMP-dependent protein kinase, *Proc. Natl. Acad. Sci. U.S.A.,* 84, 7319, 1987.

98. **Azarnia, R. and Loewenstein, W. R.,** Intercellular communication and the control of growth. X. Alteration of junctional permeability by the *src* gene. A study with temperature-sensitive mutant Rous sarcoma virus, *J. Membr. Biol.,* 82, 191, 1984.

99. **Azarnia, R. and Loewenstein, W. R.,** Polyomavirus middle T antigen downregulates junctional cell-to-cell communication, *Mol. Cell. Biol.,* 7, 946, 1987.

100. **Edelman, G. M., Cunningham, B. A., Reeke, Jr., G. N., Becker, J. W., Waxdal, M. J., and Wang, J. L.,** The covalent and three-dimensional structure of concanavalin A, *Proc. Natl. Acad. Sci. U.S.A.,* 69, 2580, 1972.

101. **Lis, H. and Sharon, N.,** Lectins as molecules and as tools, *Annu. Rev. Biochem.,* 55, 35, 1986.

102. **Gunther, G. R., Wang, J. L., Yahara, I., Cunningham, B. A., and Edelman, G. M.,** Concanavalin A derivatives with altered biological activities, *Proc. Natl. Acad. Sci. U.S.A.,* 70, 1012, 1973.

103. **Sitovsky, M. V., Pasternack, M. S., Lugo, J. P., Klein, J. R., and Eisen, H. N.,** Isolation and partial characterization of concanavalin A receptors on cloned cytotoxic T lymphocytes, *Proc. Natl. Acad. Sci. U.S.A.,* 81, 1519, 1984.

104. **Carrow, G. M. and Levitan, I. B.,** unpublished data, 1988.

105. **Kammer, K. and Burger, M. M.,** Release of cell-associated Concanavalin A by methyl α-D-mannopyranoside reveals three binding states of Concanavalin A receptors on mouse fibroblasts, *Eur. J. Biochem.,* 132, 433, 1983.

106. **Spray, D. C., Fujita, M., Saez, J. C., Choi, H., Watanabe, T., Hertzberg, E., Rosenberg, L. C., and Reid, L. M.,** Proteoglycans and glycosaminoglycans induce gap junction synthesis and function in primary liver cultures, *J. Cell Biol.,* 105, 541, 1987.

107. **Carrow, G. M. and Levitan, I. B.,** Receptor-mediated regulation of synaptic connectivity in cultured *Aplysia* neurons, *Soc. Neurosci. Abstr.,* 14, 895, 1988.

Chapter 5

HEPATOCYTE CYTODIFFERENTIATION AND CELL-TO-CELL COMMUNICATION

Michael J. Olson, Michael A. Mancini, Manjeri A. Venkatachalam, and Arun K. Roy

TABLE OF CONTENTS

I. ROLE OF LOBULAR ORGANIZATION IN THE INTEGRATION OF LIVER FUNCTION

Coordination of functional diversity with compensatory adaptability requires that the cellular constituents of a tissue or an organ develop intricate communication networks. Such functional needs also dictate that stable alterations in differential gene expression, i.e., cytodifferentiation, be part of the adaptive response. The liver is a multifunctional organ involved in the central coordination of complex metabolic and homeostatic processes. The primary functions of the mammalian liver in maintenance of homeostasis with respect to such parameters as circulating levels of glucose, synthesis and secretion of plasma proteins, and provision of complex lipids are carried out by the organ's major cell type, the parenchymal hepatocyte. Although mammalian liver is composed of about 95% parenchymal hepatocytes (80% by volume), large numbers of Kupffer cells, vascular endothelial cells, lipocytes, and bile duct epithelial cells comprise the remainder of the hepatic cell population and are increasingly recognized as essential in conditioning parenchymal cell function.

Some of the functional requirements of the liver are accommodated through a specialized vascular system and the arrangement of parenchymal and other cells as hepatic lobules or acini.[1] In the pig, the species in which lobular architecture was first described, each of the hepatic lobules is enclosed within a connective tissue capsule. However, in the case of most other mammals, such as rodents and man, there is no septal structure to demarcate hepatic lobules. The lobules appear microscopically as imperfect hexagonal prisms delimited by branches of the hepatic artery (carrying oxygenated blood) and the portal vein (carrying absorbed nutrients from the intestinal circulation). Blood flows past the parenchymal hepatocytes in modified capillary tracts known as sinusoids and is collected into terminal sinusoidal spaces, also called central veins. The central veins in turn are drained by radicles of the hepatic vein. The parenchymal hepatocytes occur as radially arranged cords or plates, originating from the lobular surface and ending at the central vein. Thus, in contrast to hepatocytes located around the central vein (pericentral), periportal hepatocytes are exposed to higher oxygen tension and substrate concentrations; distinctive enzymatic and metabolic patterns exist in these two populations of parenchymal hepatocytes to accommodate the different biochemical requirements imposed by variations in the extracellular milieu.[2-4] The concept of functional heterogeneity among periportal and pericentral hepatocytes is well established with regard to regional differences in expression of key enzymes regulating processes such as gluconeogenesis, glycolysis, and ketogenesis.[1-4] Moreover, dynamic methods used in intact, perfused livers provide convincing information that rates of these processes also vary significantly between periportal and pericentral hepatocytes.[5-7] The mechanism for differential modulation of the rates of these processes is currently undergoing active investigation but may relate to nutrient, hormonal, or other gradients established by the circulatory pattern of the liver.[3] Additional information, described in this chapter, points to regional specialization of hepatocytes for expression of proteins synthesized for export.[8-10] While these differences among parenchymal hepatocytes may be maintained by availability of critical substrates or hormones, the discovery of cell membrane regions (e.g., gap junctions) and receptors (e.g., integrins) specialized for recognition of other cells and extracellular structures, respectively, provides evidence that differentiated function is also maintained by nonhumoral mediation (reviewed in References 11 and 12).

II. FORMS OF CELLULAR INTERACTION

Within the limitations imposed by zonal differences, hepatocytes of the same cord within a lobule function in a coordinated fashion and are amply connected through gap junctions.[11] This synchronization of homologous cells is thought to be effected, in part, by exchange of low molecular weight, soluble signaling substances through the gap junctions.[11,13-19] In addition, it is becoming more and more apparent that heterologous cell-cell interactions through nonjunctional membrane contacts may also play critical roles in promoting differentiated hepatic parenchymal cell function.[20-27] Another dimension to the coordinated function of various cell types in the liver is provided by short-acting paracrine mediators secreted by one cell type to influence adjacent cells of a different type. A more elaborate extension of this type of paracrine regulation is the deposition of the extracellular matrix (ECM) at specific anatomic locations within the liver.[28] The ECM of liver contains many growth-promoting and regulatory substances which can interact with cell surface receptors, thus influencing gene expression and hepatocyte function.[12,29-36] Integration of molecular biological and cell culture methods into the study of hepatocytes has contributed greatly to a definition of the potential roles of cell-cell interaction in differentiation of parenchymal cells. However, questions about the mechanisms by which cell-cell and cell-matrix interactions selectively influence expression of the gene products associated with differentiation remain largely unanswered. In this chapter we will review our own data on hepatocellular heterogeneity of expression of the male rat protein α_{2u}-globulin in the context of the various types of intercellular signaling systems which are thought to functionally coordinate parenchymal hepatocytes.

III. ONTOGENY OF α_{2u}-GLOBULIN EXPRESSION IN MALE RAT LIVER

In sexually mature rodents several examples of regionally exclusive expression of proteins within the hepatic lobule are known.[8-10] In particular, our experience with preferential expression of proteins by parenchymal hepatocytes involves the major urinary protein identified in the mature male rat.[37] This 18-kD protein is termed α_{2u}-globulin. Parenchymal hepatocytes synthesize and secrete α_{2u}-globulin, which, due to its low molecular weight, is filtered by the renal glomerulus and excreted in urine.[37] It is not expressed until puberty (ca. 35 days of age),[38] and acquisition of competence to produce α_{2u}-globulin is primarily under androgenic control although several other hormones (e.g., growth hormone) synergistically influence α_{2u}-globulin synthesis.[39-41] Cytofluorographic and histochemical studies of the parenchymal hepatocytes of adult male rats support the existence of two populations of hepatocytes which differ in their competence to produce α_{2u}-globulin; cells in the region of the lobule adjacent to the central vein (pericentral) preferentially express the protein while periportal cells synthesize it only under conditions of maximal induction.[8,9,42,43] For the purposes of this review, then, the pericentral localization of α_{2u}-globulin synthesis in male rat liver will be a key marker of differentiated hepatocyte function.

In an attempt to understand the factors which regulate differentiation of pericentral hepatocytes for the specialized function of α_{2u}-globulin synthesis, the acquisition of competence to produce α_{2u}-globulin has been studied in male rats undergoing pu-

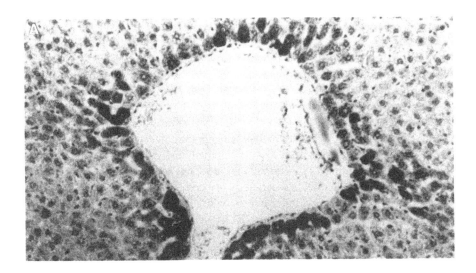

FIGURE 1. Development of competence to produce α_{2u}-globulin in parenchymal hepatocytes *in situ* during puberty in the male rat illustrating the hepatic lobular pattern of acquisition of competence to produce the protein. Sections of liver were sampled and fixed as described previously,[8] and α_{2u}-globulin was identified using an avidin-biotin immunohistochemical method with a polyclonal rabbit anti-rat-α_{2u}-globulin as the primary antibody.[8] Panel A, liver from a 40-day-old rat illustrating the presence of a unicellular layer of hepatocytes containing α_{2u}-globulin surrounding the central vein. Panels B to D show the sequence of acquisition of competence to produce α_{2u}-globulin within one hepatic lobule. Panel B, single hepatocytes immediatedly adjacent to the central vein first become capable of synthesizing α_{2u}-globulin at about day 38. In Panel C, at 40 days of age, competent cells encircle the central vein. By 42 days (Panel D), increasing numbers of cells in the pericentral region of the hepatic lobule gain the ability to produce α_{2u}-globulin. Note that radial propagation occurs only within cords of cells which possess a competent cell at the central vein. Panel E shows that cells in an adjacent noncompetent cord fail to gain competency even though in contact with competent cells (45 days of age). (From Sarker, F. H., et al., *J. Endocrinol.*, 111, 205, 1986. With permission.)

berty.[8,9] Immunochemical staining of liver sections sampled from male rats at 35 to 45 days of age reveals a specific ontogeny of developing competence to produce α_{2u}-globulin (Figure 1). At about 35 to 38 days of age, single parenchymal hepatocytes reactive with anti- α_{2u}-globulin antibody are identified immediately adjacent to the central vein of the hepatic lobule (Figure 1). These cells are in close proximity to the vascular endothelium of the central vein. By day 40, a continuous monolayer of cells competent to synthesize α_{2u}-globulin encircles the central vein suggesting lateral spread of the ability to synthesize the protein. From day 40 onward, competence is propagated radially away from the central vein through contiguous cords of hepatocytes (Figure 1). Competency progresses only within individual cords of parenchymal cells, and there is no apparent passage of the ability to produce α_{2u}-globulin to cells outside of the cord (Figure 1). *In situ* hybridization with a ^{35}S-methionine-labeled antisense cRNA homologous to α_{2u}-globulin mRNA confirms that the presence of this protein exclusively in pericentral hepatocytes is in fact due to selective derepression of the genomic segment(s) coding for α_{2u}-globulin in these cells (Figure 2). Although the capacity to synthesize α_{2u}-globulin is conferred by androgens and other hormones, the regional distribution of hepatic cells competent for α_{2u}-globulin expression appears to be under the control of factors other than hormone availability.

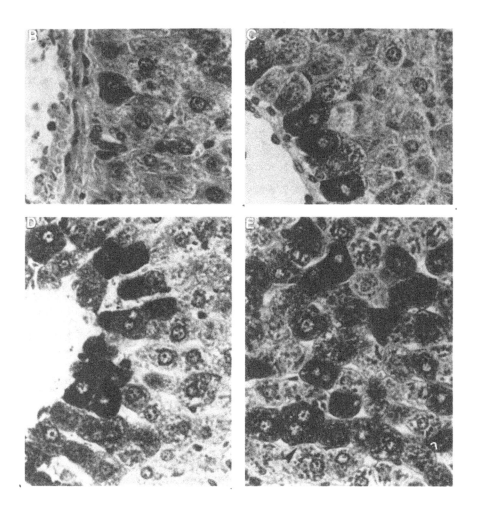

FIGURE 1 (continued)

The mechanism which is responsible for coordinating the spread of competence to produce α_{2u}-globulin in parenchymal hepatocytes is unknown. Obviously, availability of the initial hormonal signal alone can not explain this phenomenon as all hepatocytes, both periportal and pericentral, are presumably exposed to the same hormone concentration. In ovariectomized female rats administration of testosterone causes the appearance of parenchymal hepatocytes competent to produce α_{2u}-globulin in the same hepatic lobular locale (pericentral) as observed in normally developing male rat liver.[8] This observation supports the hypothesis that receptivity for the binding globulins which carry steroid hormones or androgen receptors themselves are not differentially distributed in the hepatic lobule of male rats. We have also observed that cell division, as monitored by incorporation of ^{3}H- thymidine into hepatocyte nuclei, remained low during the acquisition of the ability to synthesize α_{2u}-globulin, indicating that division of competent cells is not the mechanism for periportally directed propagation of the ability to synthesize α_{2u}-globulin.[8] Of the few cells actively incorporating thymidine

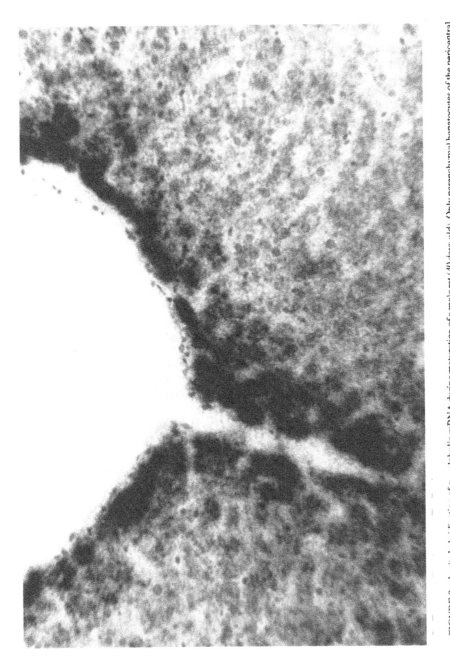

FIGURE 2. *In situ* hybridization of α_{2u}-globulin mRNA during maturation of a male rat (40 days old). Only parenchymal hepatocytes of the pericentral region bind the ^{35}S-labeled cRNA probe as shown by autoradiography of tissue sections. (From Roy, A. K. and Mancini, M. A., in *Cell-to-Cell Communication in Endocrinology*, Piva, F., Berlin, C. W., Forti, G., and Motta, M., Eds., Raven Press, New York, 1988. With permission.)

during this time, most were confined to the periportal region consistent with previously reported observations.[44] Furthermore, the *in situ* hybridization data already presented (Figure 2) allow the conclusion that whatever the nature of the differentiating influence, it is effective at the pretranslational level. Thus, with respect to production of α_{2u}-globulin, it appears that periportal and pericentral hepatocytes differ in one of only a few identifiable critical steps which occur between androgen exposure and derepression of the genomic segment(s) coding for the protein. Given these observations, a particularly interesting question arises; i.e., how is the initial hormonal stimulus for α_{2u}-globulin synthesis differentially recognized such that only specific cells of the hepatic lobule gain the ability to express this protein? To attempt an explanation, the remainder of this text explores interactions among homologous and heterologous cell types in liver as well as interactions between parenchymal hepatocytes and elements of the ECM. Figure 3, which will be referred to in the subsequent discussion, illustrates several potential mechanisms which may facilitate differential expression of α_{2u}-globulin in the central region of the hepatic lobule.

IV. ROLE OF INTERCELLULAR COMMUNICATION IN FUNCTIONAL CYTODIFFERENTIATION OF PARENCHYMAL HEPATOCYTES

A. INTERACTIONS BETWEEN NONPARENCHYMAL CELLS AND PARENCHYMAL CELLS

As mentioned previously, the development of methods for maintaining rat parenchymal hepatocytes and other liver-derived cell types under defined conditions in primary culture has allowed studies of heterologous cell-cell interactions.[20-27] Much research in this area is driven by the observation that parenchymal hepatocytes placed into primary cell culture on plastic or collagen substrates rapidly lose the ability to express a number of products or perform functions normally associated with the differentiated state,[17,18,20-27] including synthesis of α_{2u}-globulin.[40] For many specific functions, loss of activity can be reversed or the function maintained at near *in vivo* levels by inclusion of nonparenchymal hepatic cells.[20-27] Juxtaposition of cells in culture, but not direct contact, appears to be critical; conditioned medium from pure cultures of hepatic nonparenchymal cells does not stimulate differentiated function when added to primary cultures of rat hepatocytes. Further supporting this conclusion is the observation that in the isolated perfused liver, which maintains intact the various homologous and heterologous cell-cell contacts, synthesis of α_{2u}-globulin is rapidly stimulated by infusion of testosterone.[41] Important studies probing the basis for such phenomena (see Section IV.B.) demonstrate that many insoluble constituents of the ECM, produced by hepatic nonparenchymal cells, help to maintain differentiated hepatocellular function.[29-36] Furthermore, the results of primary cell culture of rat hepatocytes showing beneficial effects of either co-culture of hepatocytes with other cell types or with biological matrices allow the interpretation that these direct and indirect intercellular interactions are critical determinants of hepatocyte cytodifferentiation *in vivo*.

A well-studied example of the potential influence of heterologous cell-cell interactions on hepatic parenchymal cell differentiation, which complements our observations with α_{2u}-globulin, is the development of mouse liver and acquisition of competence to produce major urinary protein (MUP).[10] MUP is analogous to male rat α_{2u}-globulin in both molecular weight and localization of its synthesis to the liver of postpubescent mice.[10] However, unlike α_{2u}-globulin, MUP is expressed in the liver of both male and female mice.[45] Immunofluorescent detection of the binding of antibodies against MUP, glutamine synthetase, and albumin show that during adult life relatively non-specialized functions of protein production such as albumin synthesis are distributed uniformly throughout the hepatic lobule.[10] However, much like the case for male rat α_{2u}-globulin, MUP was localized exclusively to pericentral hepatocytes of mice.[10] Likewise, glutamine synthetase was identified only in pericentral parenchymal hepatocytes. Moreover, the number of cells staining for this antigen was even more limited than that reacting with the anti-MUP antibody; glutamine synthetase-positive cells occupied only a single layer surrounding the central vein.[10] Further investigation of the differential intralobular distribution of MUP production focused on primary cultures of fetal hepatoblasts, which include a variety of mesenchymally derived cell types as well as parenchymal hepatocytes.[10] The parenchymal hepatocyte-like cells of these mixed cultures were seen to undergo differentiation *in vitro* to acquire an immunostaining profile similar to adult hepatocytes. Initially, clusters of hepatic parenchymal cells derived from the fetal hepatic bud on day 13 of gestation showed uniform staining for albumin and glutamine synthetase with no evidence of MUP production. Albumin production remained constant throughout *in vitro* cytodifferentiation and occurred in all hepatocytes. After several days in culture, nearly all hepatocyte-like cells expressed glutamine synthetase but, as the culture period continued, the number of parenchymal cells producing glutamine synthetase diminished to the point that very few cells contained this antigen.[10] By day 6 of culture (equivalent to day 18 of gestation) increasing numbers of hepatocytes became competent to produce MUP and by day 12 in culture (corresponding to postpartum day 7) about 50% of the parenchymal hepatocyte-like cells were reactive with anti-MUP. Although the spatial organization into hepatic lobules is obviously lost during preparation of hepatoblast cultures, Bennett et al.[10] concluded that, in culture, the proportion of hepatocytes which were immunoreactive with the three antibodies used for screening was similar to that observed *in vivo*. Thus, the sequence of development of competence to express certain proteins *in vitro* and the proportion of cells competent for specific functions were similar to those observed during normal development *in vivo*. A controlling mechanism for the apparently programed expression of certain gene products in this situation has not been identified but the composition of the cell population derived from culture of the fetal hepatic bud provides important clues.

Fetal hepatoblasts are derived from a common epithelial precursor cell and share many common functions, such as expression of albumin, regardless of their ultimate position in the hepatic lobule. *In vitro* only 10 to 20% of the total cells isolated from the hepatic bud differentiate into parenchymal hepatocytes while the remainder form endothelial cells, fibroblasts, and other cell types.[10] Parenchymal hepatocyte-like cells in this mixed culture system are commonly found in contact with, or at least adjacent to, nonparenchymal cell types. A very similar situation was also noted during the formation of small spheroidal cellular aggregates derived from newborn rat liver in which clusters of parenchymal hepatocytes were encapsulated within a one-cell-thick layer of nonparenchymal cells.[33] Thus, in these culture systems, both of which develop or retain differentiated function *in vitro*,[10,33] heterologous (as well as homologous) intracellular interactions are maintained by chance inclusion of both parenchymal and nonparenchymal hepatic cells. On the other hand, in cultures of adult parenchymal hepatocytes where differentiated functions such as expression of MUP, α_{2u}-globulin, and other proteins are lost rapidly,[17,20-27,40] relatively few (<10%) nonparenchymal cells exist during the initial period of culture.[33]

The scope of experimentation on the role of heterologous cell-cell interaction in cytodifferentiation extends to analyses of parenchymal hepatocyte function in cultures of these cells and a companion cell type added intentionally. Such experiments have examined whether co-culture with various nonparenchymal cell types might extend the differentiated *in vivo* state of parenchymal hepatocytes in primary culture.[20-27] It appears that this is indeed the case for a growing list of functions. Initially, co-cultivation of rat parenchymal hepatocytes on feeder layers of C3H/T10^1/2 cells or fibroblasts was found to markedly lessen the rapid decline in some parenchymal hepatocyte-specific functions, such as xenobiotic metabolism and responsiveness to glucocorticoid induction of tyrosine aminotransferase activity, noted in the first few days after hepatocyte isolation.[20,21] Subsequently, improvements in the detection of microquantities of albumin and demonstration that this protein is the exclusive product of parenchymal hepatocytes led to acceptance of hepatocellular ability to synthesize albumin as a marker for retention of differentiated function *in vitro*. Guguen-Guillouzo and co-workers were among the first to report that co-culture of adult rat hepatic parenchymal cells with nonparenchymal liver cells resulted in marked improvement of cell survival and significantly extended albumin production by parenchymal hepatocytes.[22] Hepatocytes also retained characteristic polygonal morphology over periods of 8 weeks in co-culture rather than assuming the squamous appearance typical of parenchymal cells initiated into culture as a pure population on a plastic attachment substrate.[22] Furthermore, the ability of nonparenchymal cells of liver to extend differentiated hepatocyte function is not species-specific; nonparenchymal cells derived from rat liver promote extended function of human parenchymal hepatocytes in primary culture.[23] Not only does co-culture extend differentiated function but Guguen-Guillouzo et al.[22] have shown that addition of nonparenchymal cells, presumed to be derived from biliary ducts, to established cultures of rat parenchymal hepatocytes actually reversed losses in the ability to produce albumin sustained prior to addition of the nonparenchymal cells. Additional reports from the same group demonstrated further that the rate of ^{3}H-leucine incorporation into total parenchymal cell protein and also the quantity of albumin mRNA were increased by co-culture with nonparenchymal cells.[24] Assays with nuclei isolated from cultures containing both cell types showed that the measured rate of albumin mRNA transcription was increased in comparison to parenchymal cell cultures maintained without nonparenchymal companion cells.[24] This result is especially significant because studies of primary cultures of parenchymal hepatocytes alone have shown maintenance of high levels of albumin mRNA by the use of hormonally defined media. In this instance, however, post-transcriptional message stabilization rather than an increase in the rate of transcription was responsible for maintenance of high levels of albumin mRNA.[46]

At least two types of hepatic nonparenchymal cells have been shown to act cooperatively with parenchymal hepatocytes to prevent loss of differentiated function *in vitro*.[22-27] In addition to the activity of biliary epithelial cells in this regard,[22-24,26] Morin and Normand and Goulet et al. demonstrated that isolated hepatic sinusoidal endothelial cells were capable of extending the period over which cultured parenchymal hepatocytes could produce albumin at high rates and maintain the morphological appearance of hepatic parenchymal cells.[25,27] The "hepatocyte-stabilizing activity" observed with nonparenchymal cells of hepatic origin was also found to occur when pulmonary endothelial cells were co-cultured with parenchymal hepatocytes.[25] Recently, the same authors demonstrated that the beneficial effects of co-culture on maintenance of differentiated parenchymal hepatocyte function could also be imparted by continuous passage lines derived from bovine aortic endothelium, mouse embryonic fibroblasts, and human or rat dermal fibroblasts.[27] In addition to facilitating differentiated parenchymal hepatocyte function *in vitro,* sinusoidal endothelial cells proved capable of actually inducing differentiation in isolated hepatocytes. Parenchymal hepatocytes isolated from rats prior to weaning produce α-fetoprotein (AFP) in culture.[25] Co-culture with either freshly isolated sinusoidal endothelial cells or a cloned line derived from these cells caused rapid disappearance of AFP and increased expression of albumin.[25] Suppression of AFP synthesis in this culture system was interpreted as an indication of accelerated differentiation caused by the inclusion of sinusoidal lining cells.

Hepatic nonparenchymal cells also appear to facilitate replication, or at least nucleotide incorporation into DNA, of parenchymal hepatocytes. When parenchymal hepatocytes were placed in co-culture with nonparenchymal cells from rat liver, DNA synthesis was increased without addition of chemical mitogens to the culture medium.[47] Furthermore, this result was not due to soluble factors secreted into the culture medium, as conditioned medium from nonparenchymal cells, when added to hepatocyte cultures, did not stimulate thymidine incorporation into hepatocytes.[47] Cultures of nonparenchymal cells killed prior to introduction of parenchymal hepatocytes were also capable of stimulating parenchymal cell DNA synthesis. These two observations support the conclusion that nonparenchymal cells most likely support parenchymal cell replication due to formation of a substrate for parenchymal cell attachment (see Section IV.B.). Finally, only hepatic nonparenchymal cells were capable of causing this effect; continuous passage cell lines, whether of hepatic origin or not, were without effect on the rate of DNA synthesis of cultured hepatic parenchymal cells.[47]

One interpretation of the preferential expression of α_{2u}-globulin in pericentral hepatocytes which may depend on heterologous cell-cell interactions is that the initial hormonal signal for α_{2u}-globulin production may act upon vascular endothelial cells of the central vein which in turn create a secondary message that affects adjacent hepatocytes.[9] This possibility is illustrated in Figure 3. In addition, however, we have observed that hepatocytes, at the liver surface and in contact with Glisson's capsule, also express α_{2u}-globulin early in the inductive response. The fibrous capsule of the liver is rich in many of the same connective tissue components (e.g., proteoglycans) as the subendothelial space of the central vein. It is possible, then, that the cell types responsible for forming connective tissue may be primary targets of the hormones which control expression of α_{2u}-globulin (Figure 3).

Although maintenance of cell-cell contact has frequently been suggested as a principal force in regulating not only tissue morphogenesis but also in stimulating terminal differentiation of function,[48] little is known about how such actions on differentiation occur. A primary focal point of preceding chapters in this volume has been the morphology of gap junctions and function of these specialized membrane

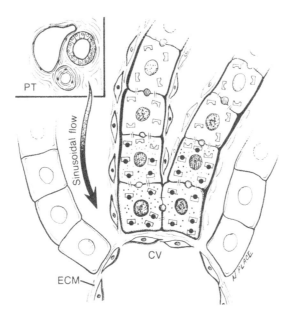

FIGURE 3. Integrated model for the possible roles of intercellular communication and effects of extracellular matrix in the hormonal induction of α_{2u}-globulin ($\cdot\cdot\therefore\cdot\therefore$) within the pericentral region of the hepatic lobule. Within the hepatic lobule, sinusoidal blood flows from the portal tract (PT) toward the central vein (CV) carrying the initial hormonal signal(s) for α_{2u}-globulin synthesis. As described in the accompanying text, vascular endothelial cells of the CV may be a primary target for the hormonal signal and, in turn, provide a second message (●) to parenchymal hepatocytes, thus initiating differential expression of α_{2u}-globulin. This second message binds to a receptor (☐) which must reach a critical level of occupancy prior to derepression of the genomic segment(s) coding for α_{2u}-globulin. Competence to synthesize the protein is then propagated away from the CV via gap junctions (⊣⊢◯⊣⊢) between parenchymal hepatocytes. Alternatively, elements of the subendothelial matrix (⇌) may bind and concentrate a signal substance released by hormonally stimulated hepatocytes "upstream" from the central lobular region and the regionally high concentration of this signal may trigger the induction of α_{2u}-globulin synthesis.

regions in communication between cells. In fact, however, preliminary evidence shows that the heterologous cell-cell interactions which prolong or restore differentiated function in cultured liver cells probably do not depend on communication mediated by gap junctions. A case in point are the results of Mesnil et al.[26] While examining the effects of co-culturing freshly isolated biliary epithelial cells or IAR 20 cells (a continuous passage line derived from biliary epithelium) with primary cultures of rat hepatic parenchymal cells, they showed that biologically significant extension of differentiated hepatocyte function (albumin production) *in vitro* was possible. However, assessment of gap junctional communication by transfer of the dye, Lucifer yellow, showed that although both IAR 20 and biliary epithelial cells were capable of homologous communication, no dye passage occurred between either of the nonparenchymal cell types and parenchymal hepatocytes.[26] Thus while the inclusion of nonparenchymal cells is critical for maintenance of differentiated parenchymal hepatocyte function *in vitro,* the likelihood that the beneficial effects of including nonparenchymal cells is brought about by gap junctional communication is low.

B. INTERACTIONS BETWEEN EXTRACELLULAR MATRIX (ECM) AND PARENCHYMAL CELLS

The foregoing section on heterologous cell-cell interactions in liver focused on the variety of cell types shown to influence parenchymal hepatocyte function *in vitro*. This discussion was facilitated by considering the effects of these cells on maintenance of representative hepatocellular functions but ignored the contribution of each nonparenchymal cell type to formation of the ECM. Increasingly sophisticated analysis of cell-specific differentiated function has led to an awareness that elements of the ECM condition the responses of hepatic parenchymal cells to a variety of stimuli.[29,31-34] Although the mechanism for the activity of ECM in this regard is as yet unknown, specific chemical components of the hepatic ECM have been identified which contact the parenchymal cell surface,[28,30,33-36,49,50] and for which there are specialized cell surface receptors (reviewed in References 12 and 51). Recently, considerations of interactions between the ECM and parenchymal hepatocytes have also examined topics such as distribution of ECM components (e.g., laminin) within the hepatic lobule,[28,36] and identified the cell types with elaborate matrix components.[30,35,36,49,50] In addition, identification and cloning of a fibronectin-laminin receptor,[52] or integrin,[12,51] points toward a role for elements of the ECM as the factor(s) supplied by nonparenchymal cells of liver which effect hepatic parenchymal cell cytodifferentiation.

The ECM of liver is concentrated primarily around the portal triad (portal vein, bile duct, and hepatic artery) and the central vein.[1,28] In these areas prominent components of the matrix include collagen types I, III, and IV, laminin, fibronectin, and several types of glycosaminoglycans and proteoglycans.[28,53,54] Unlike the regions surrounding these vascular channels, the subendothelial region in the hepatic sinusoids does not contain a clearly visible basement membrane composed of these extracellular elements.[1,34] Only recently have refined techniques of *in situ* protein immunoidentification and the availability of antibodies directed against ECM elements revealed that the region between the sinusoidal endothelial cells and parenchymal hepatocytes (the space of Disse) contains reticulin fibers comprised of collagens, laminin, and, presumably, other cellular adhesion and support molecules.[34,36,53,54] Comparison of parenchymal hepatocyte phenotype in pure culture vs. culture of hepatocytes with companion cells (either from the sinusoidal lining or bile ductules [see above]) suggests strongly that the ECM elaborated by nonparenchymal cells is responsible for not only maintenance of differentiated parenchymal hepatocyte function,[29,32-34] but also for reformation of gap junctions between freshly isolated parenchymal cells placed into primary culture with sinusoidal endothelial cells.[27] This conclusion is further supported by studies showing that individual glycosaminoglycans and proteoglycans when added to pure cultures of rat parenchymal hepatocytes increase gap junctional communication.[17,18]

The cell surface of parenchymal hepatocytes is specialized not only for passage of low molecular weight soluble substances between homologous cells (gap junctions),[11] but also contains recognition sites which bind elements of the ECM.[12,51,52] These receptors complete a chain of intercellular communication by interacting with elements of the ECM produced by adjoining cells. A number of these receptor sites recognize the amino acid sequence Arg-Gly-Asp (RGD) contained in ECM components such as collagens, vitronectin, laminin, thrombospondin, and fibronectin.[12,51,52] Aside from important effects on cell migration and positioning, interactions of parenchymal cells with elements of the ECM through specialized cell surface receptors may have profound effects on gene expression which manifest as specific differentiated hepatocyte function. Investigation of the amino acid sequence of certain chondroitin sulfate proteoglycan core proteins shows regions of homology with lectin-binding protein.[55] Therefore, it may be speculated that one mechanism whereby elements of the ECM promote parenchymal cell differentiation in liver involves binding and concentration of humoral signaling molecules in specific regions of the hepatic lobule. In addition, the chondroitin sulfate core protein shares common sequences with certain growth factors.[55] From this it may be hypothesized that matrix elements may influence cellular differentiation directly through hepatocyte integrins such as the fibronectin-laminin receptor. Perhaps, then, differentiated function in different regions of the liver lobule is regulated by variable composition of the ECM surrounding the periportal as opposed to the pericentral vascular channels. Whether this reflects variability in the proportions of different nonparenchymal cells producing the ECM in different lobular areas or is an effect of the transsinusoidal gradient of oxygen-nutrient availability on these cells is uncertain. The potential role of the ECM in promoting pericentral hepatocyte specialization for production of α_{2u}-globulin is illustrated in Figure 3.

As already described (Section IV.A.), co-culture of parenchymal hepatocytes with a variety of other cell types of hepatic and nonhepatic origin extended differentiated function which may generally be ascribed to the ability of these cells to produce components of ECM (reviewed in References 32 and 56). The results of Ben-Ze'ev et al.[29] further suggest that the specific chemical composition of the substrate or matrix used for parenchymal hepatocyte culture may be the critical determinant modulating expression of differentiated gene function. As mentioned earlier in this chapter, use of collagen promotes a flattened cell morphology in which cytoskeletal genes are abundantly expressed, this is especially apparent when cells are initiated into culture at low density.[29] On the other hand, attachment to a matrix composed of type IV collagen, heparin sulfate proteoglycan, and laminin (Matrigel®; ECM of the Engelbreth-Holm-Swarm tumor) causes isolated parenchymal hepatocytes to maintain a cuboidal morphology and preferentially express proteins and related mRNAs normally associated with hepatocyte function *in vivo*.[29,34] In contrast, the use of type I collagen alone as an attachment substrate for culture of parenchymal hepatocytes provides only a minimal improvement in maintenence of differentiated cell function compared to untreated plastic culture dishes.[57] These observations support arguments that noncollagenous components of the ECM (e.g., glycosaminoglycans, proteoglycans) are the critical effectors of differentiated hepatocyte function. The source of these essential matrix components is still somewhat controversial. However, both *in vivo* and *in vitro*, lipocytes from liver produce laminin, collagen, and sulfated glycosaminoglycans. Furthermore, parenchymal hepatocytes in culture themselves elaborate extracellular (matrix) elements such as fibrin and fibronectin,[30] while apparently contributing little to collagen synthesis.[49]

In addition to the effect of ECM and its individual components on parenchymal hepatocyte cytodifferentiation, certain exogenous chemicals are known to promote differentiation of liver cells.[58,59] The efficacy of these chemicals appears to be greatest, however, when hepatocytes are maintained on a suitable biosubstrate.[58] Freshly isolated parenchymal hepatocytes placed on culture dishes coated with type I collagen and supported in hormonally defined media fortified with dimethylsulfoxide (DMSO) express functional gene products for a variety of integral enzymatic and export proteins over a period of at least 40 days.[58] Analysis of rates of transcription for a number of these proteins showed that although all were expressed over the 40-day culture period, differential regulation of expression occurred such that rates of transcription for albumin and phosphoenolpyruvate carboxykinase were much reduced compared to intact liver while the transcriptional activities of ligandin and actin were maintained at or near *in vivo* rates.[58] However, the steady-state level of mRNA for α_{2u}-globulin was decreased to about 10% of the *in vivo* levels by day 6 in culture and further declined by day 12 to unmeasurable levels.[58] Although the loss of this particular function was attributed to lack of androgenic hormones, it may equally well reflect the fact that type I collagen, even when combined with exogenous chemical promoters of differentiation, is only partially effective in reproducing the pattern of gene expression seen in liver *in vivo*. As previously introduced, biomatrix materials composed of multiple proteinaceous and glycosaminoglycan constituents provide the most complete culture substrate yet tested for support of differentiated function.[34] In fact, liver cultures initiated by imprinting tissue explants on a plastic substrate, thus presumably transferring multiple matrix components and nonparenchymal cells as well as parenchymal hepatocytes, and cultured in a minimal medium devoid of added androgens continue to synthesize α_{2u}-globulin.[40] Thus, the chemical composition of the ECM, as well as the function of such a matrix in allowing appropriate three-dimensional orientation of parenchymal hepatocytes, appears to be critical to continued differentiated function of hepatocytes *in vitro*.

C. INTERACTIONS BETWEEN PARENCHYMAL CELLS

In contrast to the literature which describes the importance of intercellular interactions between heterologous cells in stimulating differentiated function of parenchymal hepatocytes, less is known about homologous interactions among parenchymal hepatocytes. Cellular functions which have been suggested to be mediated, at least in part, by parenchymal cell-cell interactions include the control of cell replication vs. differentiation, coordination of bile pumping through bile canaliculi, and maintenance of metabolic zonation within the hepatic lobule.

In general terms, proliferation and differentiation of hepatocytes are reciprocal functions.[60] *In vivo*, relatively few parenchymal hepatocytes (ca. 0.05%) are in the mitotic process at any time. However, the low spontaneous rate of hepatocyte division is matched to the normally limited need of the liver for replacement of differentiated cells. Stimuli such as chemically induced cell death or surgical ablation of large portions of the liver serve to induce cell replication such that the liver mass is restored (reviewed in Reference 60). Although the interplay between replication and differentiation in the liver is complex, experimental evidence suggests that cell-cell communication plays an important role in regulating liver growth during regeneration just as this phenomenon is of importance in regulation of differentiated function. Examination of the morphological and cellular structural alterations which coincide with liver growth during regeneration after hepatectomy suggest that gap junctional communication between parenchymal cells is inhibited.[61] Preliminary experiments also show that the normal pattern of α_{2u}-globulin expression within the hepatic lobule is dramatically altered following partial hepatectomy,[65] presumably due to a decrease in homologous intercellular communication required for normal expression of the protein. Moreover, stimuli for regeneration such as chemically induced hepatic necrosis,[62] or exposure of primary cultures of rat hepatocytes to toxicants,[63] cause plasma membrane alterations and perturbation of gap junction function. Whether alterations in gap junctions in these cases reflect direct toxicant-mediated damage to membrane structures or a physiological protective mechanism to prevent passage of toxicants or metabolites between cells, as suggested by Saez et al.[63] and Jongen et al.[64] is unknown. Our own results show that administration of subnecrogenic doses of hepatotoxins and tumor promoters such as dieldrin and phorbol 12-myristate-13-acetate (TPA) to male rats during the critical period of maturation in which pericentral hepatocytes gain competence to produce α_{2u}-globulin results in derangement of the normal pattern of α_{2u}- globulin synthesis within the hepatic lobule.[65] For example, rather than a regular pattern in which all pericentral hepatocytes contain immunochemically identifiable α_{2u}-globulin by 40 days of age (Figure 1), rats administered these toxins prior to puberty show only isolated groups of cells containing the protein scattered throughout the pericentral region.[65] Experimentation with primary cultures of rat and mouse hepatocytes has shown previously that dieldrin causes diminished gap junctional communication as measured by passage of [5-^{3}H]-uridine.[66] Use of the uridine transfer method, pioneered by Pitts and Sims,[67] has also shown that TPA inhibits intercellular communication between parenchymal hepatocytes of mouse.[68] We conclude that alterations in intercellular communication caused by chemical exposure and surgical ablation may be responsible for abolishing the normal distribution of α_{2u}-globulin production in liver. The point to be made is that both chemically induced toxicity and surgical reduction of liver mass result in profound changes in at least one mechanism (gap junctional communication) postulated to effect intercellular communication.[11]

Intercellular communication appears to play a role in stimulating DNA synthesis in cultured hepatocytes and by analogy in the intact liver. While this phenomenon has been investigated in heterogeneous cell populations of hepatic origin, little is known about how parenchymal hepatocytes communicate with each other to either stimulate or inhibit liver growth. The concept of contact inhibition of cell replication has for a number of years defined a general notion about the ability of cells to regulate their own growth. In early attempts to place freshly derived rat hepatocytes in primary culture it was commonly observed that after a period of substrate attachment, cell number and DNA content per culture decreased steadily as cells detached from the substrate.[20,21] The frequency with which nuclei incorporated large amounts of ^{3}H-thymidine was low, in accordance with the small proportion of hepatocytes replicating in a given period of time. Intensive work by a number of investigators showed that with *in vitro* preparations of parenchymal hepatocytes alone, incorporation of DNA precursors (^{3}H-thymidine) was stimulated maximally in low density cultures of hepatocytes by serum from partially hepatectomized rats, and submaximally by norepinephrine, glucagon, insulin, or epidermal growth factor (EGF).[69,70] Presumably the lack of extensive cell-cell contact in this case has a permissive effect in allowing the initial phases of cell division, i.e., thymidine incorporation, to commence. A recently published account of the role of cell density in expression of differentiated function suggested that cells plated at high density, which maintain a compact morphology and form pseudocords (trabeculae) *in vitro*, synthesize little DNA and express relatively low levels of cytoskeletal genes and high levels of hepatic parenchymal cell-specific genes.[29] The converse was also true, as cells at low density, which display a flattened morphology, actively synthesize DNA and effectively expressed genes coding for proteins such as actin, tubulin, vinculin, and α-actinin. That is, both the rate of DNA synthesis and expression of a variety of genes associated with differentiated hepatocyte function appear to be regulated inversely by cell population density. Thus, although the mechanism of communication between cells in co-culture during active parenchymal cell DNA synthesis is unknown, homologous cell-cell interactions appear to play a role in regulation of parenchymal cell replication. An unanswered question of primary importance is how the effects of cell-cell communication differ in promoting replication and differentiation.

Another example of interactions between parenchymal hepatocytes which is an expression of differentiated function is the coordinated pumping of bile through the bile canaliculi. Parenchymal hepatocytes synthesize and secrete bile acids and other constituents of bile as a primary function. Bile, secreted into canaliculi which lie between cords of hepatocytes is pumped toward the portal tract where it is collected in the terminal bile ducts. Coordinated pumping activity is, then, a requirement for facilitation of bile flow. Studies of the intact perfused liver demonstrate that calcium deprivation induced by perfusion with a medium nearly devoid of calcium leads to cholestasis which is mediated in part by a defect in bile salt-independent bile flow.[71] Studies by Smith et al.[15] first showed that in primary culture, parenchymal hepatocytes have rhythmic, spontaneous contraction of bile canaliculi and that this contraction is coupled such that when one canaliculus contracts, the neighboring canaliculus is likely to contract after a fixed, short interval. Further experimentation by Watanabe and Phillips in isolated hepatocytes showed that direct addition of Ca^{2+} to hepatocytes by microinjection stimulated contraction of bile canaliculi.[14] The hepatocytes in which this action of Ca^{2+} was demonstrated were in junctional communication (dye coupling) and the contractions involved both hepatocytes of the pair forming a bile canaliculus, although only one individual of the pair was injected with Ca^{2+}.[14] Furthermore, in triplet hepatocytes, which form two bile canaliculi, microinjection of a single terminal cell with Ca^{2+} results in contraction of the canaliculus formed by the injected cell and the adjacent cell and then, after a brief period, contraction of the canaliculus formed by the second and third cells of the triplet.[16] Thus, Ca^{2+} flux between coupled parenchymal hepatocytes controls bile canalicular contractility, and a similar phenomenon has been proposed to explain coordinated bile pumping in the liver of intact animals.

Expression of differentiated function by parenchymal hepatocytes is seen in the hepatic lobular pattern of certain metabolic functions. The pericentral localization of production of certain proteins (e.g., α_{2u}-globulin, MUP) has already been discussed. The sharp demarcation between lobular zones which are capable and incapable of synthesizing α_{2u}-globulin suggests that the mechanism controlling competence to produce the protein involves a threshold or trigger within the cells. Such a proposal is consistent with the existence of receptors which must reach a critical level of occupancy before α_{2u}-globulin is expressed (Figure 3). The radial propagation of α_{2u}-

globulin production could then be explained by flow of an effector or messenger species from one hepatocyte to the next through junctional complexes or by other means and would require saturation of all receptors in more centrally located cells prior to spread toward the portal tract. Such a mechanism is graphically depicted in Figure 3.

Other metabolic functions such as glycolysis, gluconeogenesis, and ketogenesis are known to operate at maximal rates only in certain parts of the hepatic lobule.[5-7] Rates of these functions are, of course, mediated by hormones but little has been offered in the way of an explanation for the heterogeneous distribution of metabolic functions within the hepatic lobule. To this time, the concept of metabolic zonation has been attributed to trans-sinusoidal gradients in O_2, nutrients, the ratio of hormones such as insulin and glucagon, and lobular differences in innervation.[2-4] Furthermore, the rapidity with which these parameters may be modulated is consistent with the knowledge that metabolically specialized zones within the liver are dynamic, increasing or contracting in size under different nutritional conditions.[2-4] These considerations, however, do not preclude a role for direct homologous cell-cell communication in maintenance of metabolic zonation. The pore size determined for mammalian gap junctions is sufficient to allow passage of small molecules (less than about 1 kDa) and many ionic species such as Ca^{2+}.[11] Although Ca^{2+} is compartmentalized within individual parenchymal hepatocytes and intracompartmental transfer is tightly regulated, ionic fluxes between cells in junctional communication have been demonstrated.[14,72] In addition, as discussed elsewhere in this volume, Ca^{2+} plays an essential role in controlling gap junctional permeability and formation. Ca^{2+} also functions in a wide variety of phosphorylation-dephosphorylation regulatory cycles which control the activity of enzymes involved in both biosynthetic and catabolic pathways (reviewed in Reference 73). Ca^{2+} acts as an intermediate messenger in processes, notably control of the rate of glycogenolysis, initiated by α-adrenergic effectors such as vasopressin and angiotensin II.[73] The protein calmodulin serves as an intracellular receptor for Ca^{2+} and participates in the regulation of the hepatic phosphorylase kinase which inactivates glycogen synthase by phosphorylation.[73] In addition, calmodulin has been shown to influence gap junction function.[74] Thus, although only speculative at this time, one mechanism for coordinated group metabolic function of hepatocytes may involve gap junctional passage of signaling molecules, such as Ca^{2+}, between cells recruited to function in a like manner.

V. CLINICAL IMPLICATIONS

The considerations above, relevant to regulated and coordinated gene expression by parenchymal hepatocytes in the normal liver also have important implications for the pathogenesis of several forms of liver disease, and the associated hepatic dysfunction seen in these disorders. Cell-cell communication is considered to be vital in the control of cellular proliferation, the regulation of synchrony and, therefore, the coordinated expression of metabolic activities in communities of cells. Thus, disordered intercellular communication could play a role in the evolution of the abnormal proliferation of hepatocytes, such as the unregulated hyperplasia seen in some chronic diseases, and in neoplasia. Vice versa, such departures from the normal state of proliferation (and differentiation) are also likely to be associated with disordered cell communication.

Abnormalities of cell-cell communication associated with uncontrolled hyperplasia may well underlie the parenchymal metabolic disturbances, dysregulated synthetic activities, and decreased detoxification processes seen in disease states such as cirrhosis of the liver. This condition is associated with disordered regenerative growth within nodules of liver parenchyma. In common alcoholic cirrhosis, degeneration of liver parenchyma affects all lobules; consequently, the fibrous tracts which replace necrotic hepatocytes span every lobule. Islands of hepatocytes between the tracts undergo regeneration, but fail to form normal lobules. In the regenerative, abnormal "nodules", central veins are absent. The mass of apparently normal regenerated hepatocytes in such cirrhotic livers may equal, if not exceed, the amount in normal liver. Nevertheless, there are manifestations of hepatic parenchymal dysfunction. It is not unreasonable to consider that some of these manifestations may be due to lack of regulation and coordination owing to disruption of normal intercellular communication. Disruption of such communication may plausibly explain such pathologic phenomena as intrahepatic cholestasis within bile canaliculi due to uncoordinated contractile activity. A number of other pathological processes in the cirrhotic liver nodules may have a similar basis.

Somewhat better is the situation in which cirrhosis occurs in livers affected by similar degrees of hepatocyte regeneration, but distributed over broader anatomic zones, in effect sparing the architecture of larger tracts of parenchyma. Regeneration may occur to comparable degrees under these circumstances, but the regenerative nodules exhibit a more normal histology. Correspondingly, dysfunction is of a lesser degree. Regeneration in these nodules may be compared to a process which restores the original mass of liver parenchyma following surgical ablation. In these regenerated livers, lobules are larger but architecture is normal.

Disordered cell-cell communication, hyperplasia and decreased cytodifferentiation may reinforce each other in repeating cycles in disease processes, and thus participate in the neoplastic transformation of hepatocytes seen in certain forms of chronic liver disease. For example, there is an increased incidence of hepatocellular carcinoma in cirrhosis of the liver, and, not uncommonly, the process is multicentric. Uncontrolled proliferative activity in regenerative nodules may well be one of the factors in the process of transformation, and abnormal intercellular communication may underlie the unregulated proliferation. The point is that, in the absence of disease, communities of hepatocytes must possess mechanisms which control proliferation and differentiation, as exemplified by the precise cessation of proliferation in surgically ablated livers after a critical mass of regenerated parenchyma is achieved. It may be suggested that these control mechanisms, based in part on intercellular communication, are interrupted by chronic diseases. Although potentially important, these considerations must remain speculative for the present. Current knowledge of even the anatomical and ultrastructural basis for disrupted cell-cell communication in liver disease is nonexistent. An understanding of normal regulation and coordination by intercellular communication will be necessary before the arduous task of unraveling the basis for altered regulation in disease can be undertaken.

ACKNOWLEDGMENT

The preliminary and published experimental work described in this chapter was supported by National Institutes of Health Grant DK-14744 to A. K. R. Work in the laboratory of M. A. V. was supported by National Institutes of Health Grant DK-37139.

REFERENCES

1. **Rappaport, A. M.,** Physioanatomic considerations, in *Diseases of the Liver*, Schiff, L. and Schiff, E. R., Eds., J.B. Lippincott, Philadelphia, 1978, 1.
2. **Jungermann, K. and Katz, N.,** Functional hepatocellular heterogeneity, *Hepatology*, 2, 385, 1982.
3. **Jungermann, K.,** Zonal signal heterogeneity and induction of hepatocyte heterogeneity, in *Regulation of Hepatic Metabolism: Intra- and Intercellular Compartmentation*, Thurman, R. G., Kauffman, F. C., and Jungermann, K., Eds., Plenum Press, New York, 1986, 445.
4. **Gumucio, J. J. and Miller, D. L.,** Liver cell heterogeneity, in *The Liver: Biology and Pathobiology*, Arias, I., Popper, H., Schacter, D., and Schafritz, D. A., Eds., Raven Press, New York, 1982, 647.
5. **Matsumura, T., Kashiwagi, T., Meren, H., and Thurman, R. G.,** Gluconeogenesis predominates in the periportal region of the liver lobule, *Eur. J. Biochem.*, 144, 409, 1984.
6. **Matsumura, T. and Thurman, R. G.,** Predominance of glycolysis in pericentral regions of the liver lobule, *Eur. J. Biochem.*, 140, 229, 1984.
7. **Olson, M. J. and Thurman, R. G.,** Quantitation of ketogenesis in periportal and pericentral regions of the liver lobule, *Arch. Biochem. Biophys.*, 253, 26, 1987.
8. **Sarkar, F. H., Mancini, M A., Nag, A. C., and Roy, A. K.,** Cellular interactions in the hormonal induction of α_{2u}-globulin in rat liver, *J. Endocrinol.*, 111, 205, 1986.
9. **Roy, A. K., Sarkar, F. H., Nag, A. C., and Mancini, M. A.,** Role of cytodifferentiation and cell-cell communication in the androgen dependent expression of α_{2u}-globulin gene in rat liver, in *Cellular Endocrinology: Hormonal Control of Embryonic and Cellular Differentiation*, Serrero, G. and Hayashi, J., Eds., Alan R. Liss, New York, 1986, 401.
10. **Bennett, A. L., Paulson, K. E., Miller, R. E., and Darnell, J. E., Jr.,** Acquisition of antigens characteristic of adult pericentral hepatocytes by differentiating fetal hepatoblasts *in vitro, J. Cell Biol.*, 105, 1073, 1987.
11. **Gilula, N. B. and Hertzberg, E. L.,** Communication and gap junctions, in *The Liver: Biology and Pathobiology*, Arias, I., Popper, H., Schacter, D., and Schafritz, D. A., Eds., Raven Press, New York, 1982, 615.
12. **Hynes, R. O.,** Integrins: a family of cell surface receptors, *Cell*, 48, 549, 1987.
13. **Nakamura, T., Yoshimoto, K., Nakayama, Y., Tomita, Y., and Ichihara, A.,** Reciprocal modulation of growth and differentiated functions of mature rat hepatocytes in primary culture by cell-cell contact and cell membranes, *Proc. Natl. Acad. Sci. U.S.A.*, 80, 7229, 1983.
14. **Watanabe, S., and Phillips, M. J.,** Ca^{2+} causes active contraction of bile canaliculi: direct evidence from microinjection studies, *Proc. Natl. Acad. Sci. U.S.A.*, 81, 6164, 1984.
15. **Smith, C. R., Oshio, C., Miyairi, M., Katz, H., and Phillips, M. J.,** Coordination of contractile activity of bile canaliculi. Evidence from spontaneous contractions *in vitro, Lab. Invest.*, 53, 270, 1985.
16. **Watanabe, S., Smith, C. R., and Phillips, M. J.,** Coordination of the contractile activity of bile canaliculi. Evidence from calcium microinjection of triplet hepatocytes, *Lab. Invest.*, 53, 275, 1985.
17. **Fujita, M., Spray, D. C., Choi, H., Saez, J. C., Watanabe, T., Rosenberg, L. C., Hertzberg, E. L., and Reid, L. M.,** Glycosaminoglycans and proteoglycans induce gap junction expression and restore transcription of tissue-specific mRNAs in primary liver cultures, *Hepatology*, 7, 1S, 1987.
18. **Spray, D. C., Fujita, M., Saez, J. C., Choi, H., Watanabe, T., Hertzberg, E., Rosenberg, L. C., and Reid, L. M.,** Proteoglycans and glycosaminoglycans induce gap junction synthesis and function in primary liver cultures, *J. Cell Biol.*, 105, 541, 1987.
19. **Casteleijn, E., Kuiper, J., Van Roolj, H. -C., Kamps, `J. -A., Koster, J. -F., and Van Berkel, T. -J.,** Endotoxin stimulates glycogenolysis in the liver by means of intercellular communication, *J. Biol. Chem.*, 263, 6953, 1988.
20. **Langenbach, R., Malick, L., Tompa, A., Kuszynski, C., Freed, H., and Huberman, E.,** Maintenance of adult rat hepatocytes on C3H/10T½ cells, *Cancer Res.*, 39, 3509, 1979.
21. **Michalopoulos, G., Russell, F., and Biles, C.,** Primary cultures of hepatocytes on human fibroblasts, *In Vitro*, 15, 796, 1979.
22. **Guguen-Guillouzo, C., Clement, B., Baffet, G., Beamont, C., Morel-Chany, E., Glaise, D., and Guillouzo, A.,** Maintenance and reversibility of active albumin secretion by adult rat hepatocytes co-cultured with another liver epithelial cell type, *Exp. Cell Res.*, 143, 47, 1983.

23. **Clement, B., Guguen-Guillouzo, C., Campion, J.-P., Glaise, D., Bourel, M., and Guillouzo, A.,** Long-term co-cultures of adult human hepatocytes with rat liver epithelial cells: modulation of albumin secretion and accumulation of extracellular material, *Hepatology*, 4, 373, 1984.

24. **Fraslin, J. -M., Knelp, B., Vaulont, S., Glaise, D., Munnich, A., and Guguen-Guillouzo, C.,** Dependence of hepatocyte-specific gene expression on cell-cell interactions in primary culture, *EMBO J.*, 4, 2487, 1985.

25. **Morin, O. and Normand, C.,** Long-term maintenance of hepatocyte functional activity in co-culture: requirements for sinusoidal endothelial cells and dexamethasone, *J. Cell. Physiol.*, 129, 103, 1986.

26. **Mesnil, M., Fraslin, J. -M., Piccoli, C., Yamasaki, H., and Guguen-Guillouzo, C.,** Cell contact but not gap junctional communication (dye coupling) with biliary epithelial cells is required for hepatocytes to maintain differentiated functions, *Exp. Cell Res.*, 173, 524, 1987.

27. **Goulet, F., Normand, C., and Morin, O.,** Cellular interactions promote tissue-specific function, biomatrix deposition and junctional communication in primary cultured hepatocytes, *Hepatology*, 8, 1010, 1988.

28. **Rojkind, M. and Ponce-Noyola, P.,** The extracellular matrix of liver, *Coll. Relat. Res.*, 2, 151, 1982.

29. **Ben-Ze'ev, A., Robinson, G. S., Bucher, N. L. R., and Farmer, S. R.,** Cell-cell and cell-matrix interactions differentially regulate the expression of hepatic and cytoskeletal genes in primary cultures of rat hepatocytes, *Proc. Natl. Acad. Sci. U.S.A.*, 85, 2161, 1988.

30. **Stamatoglou, S. C., Hughes, R. C., and Lindahl, U.,** Rat hepatocytes in serum-free primary culture elaborate an extensive extracellular matrix containing fibrin and fibronectin, *J. Cell Biol.*, 105, 2417, 1987.

31. **Enat, R., Jefferson, D. M., Ruiz-Opazo, N., Gatmaitan, Z., Leinwand, L. A., and Reid, L M.,** Hepatocyte proliferation *in vitro*: its dependence on the use of serum-free hormonally defined medium and substrata of extracellular matrix, *Proc. Natl. Acad. Sci. U.S.A.*, 81, 1411, 1984.

32. **Reid, L. M. and Jefferson, D. M.,** Culturing hepatocytes and other differentiated cells, *Hepatology*, 4, 548, 1984.

33. **Landry, J., Bernier, D., Ouellet, C., Goyette, R., and Marceau, N.,** Spheroidal aggregate culture of rat liver cells: histiotypic reorganization, biomatrix deposition, and maintenance of functional activities, *J. Cell Biol.*, 101, 914, 1985.

34. **Bissell, D. M., Arenson, D. M., Maher, J. J., and Roll, F. J.,** Support of cultured hepatocytes by a laminin-rich gel: evidence for a functionally significant subendothelial matrix in normal liver, *J. Clin. Invest.*, 79, 801, 1987.

35. **Arenson, D. M., Friedman, S. L., and Bissell, D. M.,** Formation of extracellular matrix in normal rat liver: lipocytes as a major source of proteoglycan, *Gastroenterology*, 95, 441, 1988.

36. **Maher, J. J., Friedman, S. L., Roll, J. R., and Bissell, D. M.,** Immunolocalization of laminin in normal rat liver and biosynthesis of laminin by hepatic lipocytes in primary culture, *Gastroenterology*, 94, 1053, 1988.

37. **Roy, A. K., Neuhaus, O. W., and Harmison, C. R.,** Preparation and characterization of a sex-dependent rat urinary protein, *Biochim. Biophys. Acta*, 127, 72, 1966.

38. **Roy, A. K., Chatterjee, B., Demyan, W. F., Milin, B. S., Motwani, N. M., Nath, T. S., and Schiop, M. J.,** Hormone and age-dependent regulation of α_{2u}-globulin gene expression, *Recent Prog. Horm. Res.*, 39, 425, 1983.

39. **Roy, A. K.,** Hormonal regulation of α_{2u}-globulin synthesis in rat liver, *Biochem. Act. Horm.*, 6, 481, 1979.

40. **Motwani, N. M., Unaker, N. J., and Roy, A. K.,** Multiple hormone requirement for the synthesis of α_{2u}-globulin by monolayers of rat hepatocytes in long term primary culture, *Endocrinology*, 107, 1606, 1980.

41. **Murty, C. V. R., Rao, K. V. S., and Roy, A. K.,** Independent regulatory influence of androgen and growth hormone on the hepatic synthesis of α_{2u}-globulin, *Endocrinology*, 116, 109, 1985.

42. **Antakly, T., Lynch, K. R., Nakhasi, H. L., and Feigelson, P.,** Cellular dynamics of the hormonal and developmental induction of hepatic alpha- 2u-globulin as demonstrated by immunocytochemistry and specific mRNA monitoring, *Am. J. Anat.*, 165, 211, 1982.

43. **Motwani, N. M., Caron, D., Demyan, W. F., Chatterjee, B., Hunter, S., Poulik, M. D., and Roy, A. K.,** Monoclonal antibodies to α_{2u}-globulin and immunocytofluorometric analysis of α_{2u}-globulin synthesizing hepatocytes during androgenic induction and aging, *J. Biol. Chem.*, 259, 3653, 1984.

44. **Grisham, J.W.**, Lobular distribution of hepatic nuclei labeled with tritiated-thymidine in partially hepatectomized rats, *Fed. Proc.*, 18, 478, 1959.

45. **Clissold, P. M., Hainey, S., and Bishop, J. O.**, Messenger RNAs coding for mouse major urinary proteins are differentially induced by testosterone, *Biochem. Genet.*, 22, 379, 1984.

46. **Jefferson, D. M., Clayton, D. F., Darnell, J. E., and Reid, L. M.**, Post-transcriptional modulation of gene expression in cultured rat hepatocytes, *Mol. Cell Biol.*, 4, 1929, 1984.

47. **Shimaoka, S., Nakamura, T., and Ichihara, A.**, Stimulation of growth of primary cultured adult rat hepatocytes without growth factors by coculture with nonparenchymal liver cells, *Exp. Cell Res.*, 172, 228, 1987.

48. **Glaser, L.**, Cell recognition: phenomena in search of molecules, in *Cellular Recognition*, Frazier, W. A., Ed., Alan R. Liss, New York, 1982, 759.

49. **Maher, J. J., Bissell, D. M., Friedman, S. L., and Roll, J. F.**, Collagen measured in primary cultures of normal rat hepatocytes derives from lipocytes within the monolayer, *J. Clin. Invest.*, 82, 450, 1988.

50. **Schafer, S., Zerbe, O., and Gressner, A. M.**, The synthesis of proteoglycans in fat-storing cells of rat liver, *Hepatology*, 7, 680, 1987.

51. **Ruoslahti, E. and Pierschbacher, M. D.**, New perspectives in cell adhesion: RGD and integrins, *Science*, 238, 491, 1987.

52. **Horwitz, A., Duggin, K., Greggs, R., Decker, C., and Buck, C.**, The cell substrate attachment (CSAT) antigen has properties of a receptor for laminin and fibronectin, *J. Cell Biol.*, 101, 2134, 1985.

53. **Hahn, E., Wick, G., Pencev, D., and Timpl, R.**, Distribution of basement membrane proteins in normal and fibrotic human liver: collagen type IV, laminin, and fibronectin, *Gut*, 21, 63, 1980.

54. **Geerts, A., Geuze, H. J., Slot, J. -W., Voss, B., Schuppan, D., Schellinck, P., and Wisse, E.**, Immunogold localization of procollagen III, fibronectin and heparin sulfate proteoglycan on ultrathin frozen section of normal liver, *Histochemistry*, 84, 355, 1986.

55. **Krusius, T., Gehlsen, K. R., and Ruoslahti, E.**, A fibroblast chondroitin sulfate proteoglycan core protein contains lectin-like and growth factor-like sequences, *J. Biol. Chem.*, 262, 13120, 1987.

56. **Maher, J. J.**, Primary hepatocyte culture: is it home away from home?, *Hepatology*, 8, 1162, 1988.

57. **Bissell, D. M. and Guzelian, P. S.**, Phenotypic stability of adult rat hepatocytes in primary monolayer culture, *Ann. N.Y. Acad. Sci.*, 349, 85, 1980.

58. **Isom, H., Georgoff, I., Salditt-Georgoff, M., and Darnell, J. E., Jr.**, Persistence of liver-specific messenger RNA in cultured hepatocytes: different regulatory events for different genes, *J. Cell Biol.*, 105, 2878, 1987.

59. **Miyazaki, M., Handa, Y., Oda, M.**, Long-term survival of functional hepatocytes from adult rat in the presence of phenobarbital in primary culture, *Exp. Cell Res.*, 159, 176, 1985.

60. **Leffert, H. L., Koch, K. S., Lad, P. J., Skelly, H., and deHemptinne, B.**, Hepatocyte regeneration, replication, and differentiation, in *The Liver: Biology and Pathobiology*, Arias, I., Popper, H., Schacter, D., and Schafritz, D. A., Eds., Raven Press, New York, 1982, 601.

61. **Dermietzel, R., Yancey, S. B., Traub, O., Willecke, K., and Revel, J. -P.**, Major loss of the 28-kD protein of gap junctions in proliferating hepatocytes, *J. Cell Biol.*, 105, 1925, 1987.

62. **James, J. L., Friend, D. S., MacDonald, J. R., and Smuckler, E. A.**, Alterations in hepatocyte plasma membrane in carbon tetrachloride poisoning. Freeze-fracture analysis of gap junctions and electron spin resonance analysis of lipid fluidity, *Lab. Invest.*, 54, 268, 1986.

63. **Saez, J. C., Bennett, M. V. L., and Spray, D. C.**, Carbon tetrachloride at hepatotoxic levels blocks reversibly gap junctions between rat hepatocytes, *Science*, 236, 967, 1987.

64. **Jongen, W. M. F., Van Der Leede, B. J. N., Chang, C. C., and Trosko, J. E.**, The transport of reactive intermediates in a co-cultivation system: the role of intercellular communication, *Carcinogenesis*, 8, 1239, 1987.

65. **Mancini, M. A., Olson, M. J., and Roy, A. K.**, unpublished observations.

66. **Klaunig, J. E. and Ruch, R. J.**, Strain and species effects on the inhibition of hepatocyte intercellular communication by liver tumor promoters, *Cancer Lett.*, 36, 161, 1987.

67. **Pitts, J. D. and Sims, J. W.**, Permeability of junctions between animal cells: intercellular transfer of nucleotides but not macromolecules, *Exp. Cell Res.*, 104, 153, 1977.

68. **Ruch, R. J., Klaunig, J. E., and Pereira, M. A.**, Inhibition of intercellular communication between mouse hepatocytes by tumor promoters, *Toxicol. Appl. Pharmacol.*, 87, 111, 1987.

69. **Michalopoulos, G., Cianciulli, H. D., Novotny, A. R., Klingerman, A. D., Strom, S. C., and Jirtle, R. L.,** Liver regeneration studies with rat hepatocytes in primary culture, *Cancer Res.,* 42, 4673, 1982.

70. **Cruise, J. L. and Michalopoulos, G.,** Norepinephrine and epidermal growth factor: dynamics of their interaction in the stimulation of hepatocyte DNA synthesis, *J. Cell. Physiol.,* 125, 45, 1985.

71. **Reichen, J., Berr, F., Le, M., and Warren, G. H.,** Characterization of calcium deprivation-induced cholestasis in the perfused rat liver, *Am. J. Physiol.,* 249, G48, 1985.

72. **Gilula, N. B., Reeves, O. R., and Steinbach, A.,** Metabolic coupling, ionic coupling and cell contacts, *Nature,* 235, 262, 1972.

73. **Williamson, J. R., Cooper, R. H., and Hoek, J. B.,** Role of calcium in the hormonal regulation of liver metabolism, *Biochim. Biophys. Acta,* 639, 243, 1981.

74. **Peracchia, C. and Girsch, S. J.,** Functional modulation of cell coupling: evidence for a calmodulin-driven channel gate, *Am. J. Physiol.,* 248, H765, 1985.

Chapter 6

JUNCTIONAL COMMUNICATION AND THE WOUND HEALING RESPONSE

David M. Larson

TABLE OF CONTENTS

I. INTRODUCTION

Writing a chapter with the title "Junctional Communication and the Wound-Healing Response" initially seemed rather presumptuous. First, while much is known about the multiplicity of events occurring during wound healing (for reviews see Reference 1), very little is known directly about any potential roles of junctional transfer in these processes. This paucity of direct information is chiefly due to the fact that few direct studies of transfer in wound healing (and none of actual communication) have been carried out.

Nevertheless, wounds are everyday facts of life; wound healing is a common biological response to a variety of insults, and perturbations of a homeostatic situation are a cornerstone of scientific methodology. Hence, there is the potential in this natural response for elucidation of the roles of intercellular interactions, including those mediated by gap junctions.

The first purpose of this review is to summarize what is known about any relationships between gap junctions, junctional transfer and junctional communication and wound healing. The second purpose is to point out different possible roles for junctional communication and to suggest directions for future studies.

A. JUNCTIONAL COMMUNICATION: DEFINITIONS

For the purpose of this review, the following terms should be defined (or redefined): gap junctions, junctional transfer, and junctional communication. *Gap junctions* are ultrastructurally defined specializations of the plasma membranes of adjacent cells.[2-5] They consist of (usually macular) assemblages of membrane channels and are apparently the site of direct cytoplasmic continuity between cells. In vertebrates, these channels have an upper permeability limit to ions and small molecules of approximately 1000 D.[6] However junctional permeability is subject to complex regulation by a variety of factors (including voltage, H^+, Ca^{2+}, cAMP; see Chapter 1) so ultrastructural detection of gap junctions, while suggestive, does not guarantee junctional transfer. *Junctional transfer* is the passage of some, known or unknown, ionic or molecular species from one cell to another via gap junctional channels. Current flow through junctions (electrotonic coupling), dye transfer, and metabolic cooperation are all examples of junctional transfer. These phenomena are what researchers commonly assay.[7] *Junctional communication* is the physiological counterpart of junctional transfer. For true communication to occur, there must be, at least, a source (transmitter) of a message, message content, a means of transmission, a receiver, a detector, and (at least potentially) a response. In the field of gap junctional communication, with the exception of electrotonic synapses in excitable cells[8,9] and a few other examples, we know precious little about most of these components. True, any cells linked by gap junctions (the means of transmission) potentially constitute transmitters and receivers. However, the potential messages (molecules or ions) and the means of detection of message are primarily speculative.

B. ROLES OF JUNCTIONAL TRANSFER

Two general and related roles of junctional transfer, relevant to wound healing, seem fairly unassailable, although again hard evidence is difficult to obtain. First, it seems obvious that junctional transfer has to subserve a homeostatic role, maintaining the intracellular concentration of small molecules and ions in a relative balance throughout a population of coupled cells.[10-12] Second, it seems clear that alterations in this homeostasis, caused by extrinsic or intrinsic forces, would be integrated to some degree (depending on electrochemical gradients, resistivities, diffusion coefficients, and junctional permeabilities and conductances) throughout such a population by junctional transfer.[10,13,14] This latter function is the mechanism of integration of contractile function in cardiac[15,16] and smooth muscle[17-19] (see Chapter 3) and of secretory events in the exocrine pancreas.[20,21] These roles, homeostasis and integration of tissue responses, are therefore reasonably secure.

One further postulated role of junctional transfer, of critical importance to wound healing, has been in the control of growth.[22-25] This hypothesis is attractive but has been difficult to test critically. Various studies on junctional transfer and ultrastructure in transformed cell lines have suggested a general negative correlation between growth control and the existence or extent of gap junctions and junctional transfer.[22] Such correlative studies provide pointers toward a link between growth control and the extent of junctions and junctional transfer (see Chapters 7 and 8). Other examples of possible relationships are discussed in this review.

II. INTERCELLULAR COMMUNICATION AND WOUNDING

A. WOUNDING AND DAMAGE CONTROL

Traditionally, a distinction is made in healing between wounds with clean, apposed edges (healing by primary intention, as of a scalpel incision) and wounds with separated edges (healing by secondary intention, as of a gouged wound, where tissue is removed). However, this distinction is quantitative (time required for healing, amount of scar tissue, etc.) rather than qualitative.

Whatever the type of wound, the primary initial event in wounding a tissue is damage to cells. At the edge of the wound, cell membranes are disrupted, leading to loss of cellular homeostasis and ionic gradients, and, potentially, death of the cell. In a population of cells linked by gap junctions, loss of membrane integrity in part of the population could result in damage or death of all the cells by a gradual loss of gradients through junctional transfer. This does not happen in coupled populations because the junctional channels close, sealing off the damaged cells. This closure is probably due a massive influx of calcium ions through the disrupted membranes and gating of the junctional channels by elevated intracellular free calcium.[26,27] For example, cut-end loading of cardiac muscle bundles in the presence of extracellular calcium labels only the damaged cells, whereas carrying out the same experiment in the absence of extracellular calcium results in junctional transfer along the bundle to undamaged cells.[28] Hence, in damaged tissue, gating of junctional transfer results in a limitation to the area of cellular mortality.

This protective gating is neither immediate nor necessarily irreversible; a recently introduced junctional transfer assay (scrape-loading)[29,30] takes advantage of the ability of damaged cells to reseal and retain the capability for junctional transfer. Cells damaged in the presence of the fluorescent dye Lucifer yellow CH take up the dye and then reseal. Transfer to adjacent, undamaged cells can be detected within seconds, indicating the presence of patent junctions.[30] These experiments demonstrate that junctional transfer is not lost due to calcium influx if membrane integrity and calcium gradients are reestablished.

B. WOUND SIGNALS?

Studies on wounding vascular endothelial monolayers and epithelial sheets, *in vivo* and *in vitro,* have shown that cellular motile and mitotic activities are increased in cells far removed from the wound edge.[31-34] These results suggest that a wound signal is generated and affects the population responses of the remaining, uninjured cells. A series of studies by Hollenberg and Odori[35] and Hollenberg et al.,[36] using a variety of injuries and insults to vessels, have suggested that a wave of endothelial mitoses travels up- and downstream from the site of injury. Similarly, studies on *in vivo* denuding endothelial wounds in vessels, in which activation of uninjured cells occurred both above and below the wound site,[31,32] make an extracellular, soluble growth or chemotactic/chemokinetic factor an unlikely mediator of this response. An intriguing possibility is junctional transfer of a wound signal. Although the nature of such a signal is unknown, a recent preliminary report by Atkinson[37] may shed some light. In this study, the microelectrode impalement of a cultured NRK cell led to a transient increase in intracellular free calcium (detected using fura-2) in that cell, followed by calcium transients in adjacent cells. Strong depolarization of a cell by current injection also led to calcium transients, first in the injected cell, then in adjacent cells. It is certainly possible that a wounded cell could transmit a similar signal (whether in the form of Ca^{2+}, second messengers, or current) to adjacent cells. If such a signal were regenerative, due for instance to calcium channel activity,[37] then it could presumably be transmitted throughout a large population of cultured cells, resulting in a general activation of uninjured cells. While this scenario is highly speculative, it provides a useful, testable hypothesis for an actual example of junctional communication.

III. INTERCELLULAR COMMUNICATION AND WOUND HEALING

For the purposes of this review, three frequently cited models of wound healing will be focused on: skin (as an example of the full spectrum of wound healing responses) and hepatocytes and vascular endothelium (as models of regeneration). The normal distribution of gap junctions and the extent of junctional transfer has been studied in each of these systems and provides the bases for comparison with the wounds.

A. JUNCTIONS AND TRANSFER IN MODEL SYSTEMS

In skin, the nucleated cells of the stratified squamous epithelium have been shown to be linked by gap junctions[38-40] although other reports have demonstrated few gap junctions,[41,42] and elaboration of gap junctions in various epithelia may be a function of cell type and developmental stage.[43,44] White et al.,[38] in studies on hamster cheek pouch epithelium, reported a considerably greater gap junctional area between cells in the stratum spinosum as compared to the stratum basale and stratum granulosum. Kam et al.[45] studied the distribution of intracellularly injected Lucifer yellow in newborn mouse skin. They reported extensive dye transfer in the dermis, less transfer in the epidermis, and restricted transfer between the two layers. In followup work on cultured keratinocytes,[46] they found a decrease in dye transfer associated with terminal differentiation (involucrin-positive cells) as compared to the more undifferentiated keratinocytes. Taken together, these data demonstrate that normal mammalian epidermal and dermal cells have gap junctions and are capable of junctional transfer, although to a variable degree depending on cell type, stage of differentiation, and the age of the animal.

In liver, the hepatocytes are richly endowed with gap junctions (an average of 3% of the contact area between adjacent hepatocytes[47-50] in rat liver). Dye transfer is extensive in normal rat liver[49,51] and both rat[49] and mouse[52] hepatocytes have been shown to be electrotonically coupled *in situ*. The extent of hepatocyte junctions and junctional transfer throughout uninjured rat liver apparently does not vary significantly.[49]

The distribution of gap junctions in vascular endothelium depends on the vessel type. Ultrastructural analyses by Simionescu and Simionescu[53] and others[54-58] have demonstrated large, frequent gap junctions between arteriolar endothelial cells, somewhat reduced junctions between arterial endothelium, and further reduced junctions in veins.[54,55,57-61] Although the several groups have been unable to detect gap junctions between endothelial cells in capillaries or pericytic venules,[54,62,64] there are scattered reports that they do exist.[65-68] Junctional transfer between endothelial cells *in situ* has not been entirely characterized, although dye transfer between endothelial cells in rat omental[69,70] and bovine brain and retinal[70-72] microvessels, including capillaries, has been reported. Dye transfer between rat aortic endothelial cells *in situ* has also been demonstrated[72] (Larson, unpublished observations). Therefore, normal vascular endothelial cells have gap junctions and engage in junctional transfer *in vivo*. Gap junctions and junctional transfer can also be demonstrated in cultured endothelium.[30,70-76]

For each of the above models, there are suggestive data for alterations in the size or frequency of gap junctions or the extent of junctional transfer in wounded tissues and culture models, as discussed below.

B. WOUND HEALING

Wound healing can be conveniently subdivided into four phases: inflammation, contraction, repair (formation of granulation tissue), and regeneration. These phases are not clearly separable in time since granulation tissue formation e.g., begins during the inflammatory phase, as does regeneration. The major characteristics of each phase are outlined to orient the reader unfamiliar with wound healing; evidence for an involvement of intercellular communication in each phase is then considered.

1. Inflammatory Phase

In any vascularized tissue, wounding usually results in damage to blood vessels, resulting in extravasation of plasma and blood cells.[78,79] The immediate result of loss of vascular integrity is platelet activation and degranulation, coagulation (fibrin deposition), and release of a variety of soluble factors. Clot formation in the wound serves a protective function, to limit loss of fluids and to stabilize the wounded area. In addition, clotting within the damaged vessels serves to delimit the area of damage and to prevent further loss of plasma constituents. Soluble mediators released at the wound site (from coagulation products, cellular activation, and cell debris) induce inflammatory cell accumulation and increase vascular permeability in the surrounding, undamaged vessels, thereby increasing the reaction. Neutrophils are the first inflammatory cells to invade the wound site. They are phagocytic, and therefore help to clear debris, but they also often exacerbate tissue damage by release of, e.g., toxic oxygen metabolites. Peripheral blood monocytes accumulate in the wound region, are activated, and take on the characteristics of macrophages. Macrophages, in addition to their phagocytic role, release a wide variety of chemoattractant and growth factors,[80] which are necessary for the formation of granulation tissue.

Although there is no immediately obvious junctional involvement in the processes mentioned above, speculative points may be made. First, the release of soluble mediators especially from platelets and macrophages, and the release of a variety of substances from necrotic cells, could have effects on gap junctions and junctional transfer throughout the surrounding tissue. A possible scenario[25,81] would be release of platelet-derived growth factor from degranulating platelets, followed by receptor occupation on a variety of responsive cell types, followed by production of IP$_3$ and diacylglycerol (DAG), leading to release of free calcium from intracellular stores and activation of protein kinase C, and possibly, phosphorylation of junctional proteins. Both the changes in $[Ca^{2+}]_i$ and phosphorylation could act to alter junctional transfer, upsetting the homeostatic control of the cells and, possibly, initiating mitotic activity[25] or allowing release of cells for migration. Similar effects might be initiated by the elaboration of other growth and chemoattractant factors. All of these effects would be felt in the following stages of wound healing.

Although there is no direct evidence for an involvement of junctional transfer in inflammation *per se*, there is some scanty evidence that emigration of leukocytes might involve transient junction formation between these cells and vascular endothelium.[82,83] Such electron microscopic evidence is certainly controversial but, at least in one culture system, possible dye transfer has also been detected between emigrating lymphocytes and endothelial monolayers.[84] The purpose of this possible transfer capability, whether it also applies to other leukocytes, and whether it may be modulated in a wound healing situation are unknown at present.

2. Contraction

This phase of wound healing is a mechanical reduction in the size of the wound primarily as a result of active contraction by myofibroblasts and is prominent in healing of skin and gastrointestinal and genitourinary tract wounds.[85] Under some circumstances, contraction can decrease the size of a wound by 70%, leading to faster healing. The myofibroblasts that primarily mediate this contraction are thought to derive from tissue fibroblasts (although vascular pericytes or mesenchymal stem cells are also possibilities)[85,86] and migrate into the wound area under the influence of chemotactic factors released during the inflammatory reaction.[85] Evidence has been presented for contraction by epithelium as well, a "purse-string" effect, since it can be shown that some contraction of wounds can occur before granulation tissue (containing myofibroblasts) forms.[87]

The role of junctional transfer in wound contraction was espoused by Ryan et al., Gabbiani et al., and Gabbiani and Rungger-Brändle in their studies on myofibroblasts.[88-90] Normal tissue fibroblasts are poorly differentiated cells that contain little organized cytoskeletal elements and apparently lack gap junctions.[90] When they "modulate" into myofibroblasts, they increase cytoskeletal organization (increased microfilaments and myosin) and develop gap junctions at cell contact points.[85] Hence, in some ways myofibroblasts are more similar to the familiar tissue culture fibroblast than the tissue fibroblast. Since the contraction of wounds *in vivo,* as well as collagen gels *in vitro,*[91] appears to be carried out by these cells in a coordinated fashion, it is natural to suggest that junctional communication is involved in the coordination. One interesting study[92] demonstrating the contraction of collagen gels by cultured fibroblasts showed numerous gap junctions between the cells during the early, active phase of contraction. Once the gel was contracted, these fibroblasts relaxed and lost their gap junctions.

Similar hypotheses have been proposed for the coordination of epithelial cells in wound contraction.[43,93] However, the evidence for a junctionally coordinated synchronization of contraction in all of these models remains correlative and circumstantial. To my knowledge, no prospective studies on these phenomena have been carried out to critically test for the involvement of junctional transfer.

3. Repair (Granulation Tissue Formation and Resolution)

The early aspect of this phase of the wound repair scheme includes the accumulation of macrophages, fibroblasts (fibroplasia), new blood vessels (angiogenesis), and loose connective tissue at the wound site. Fibroblasts in the granulation tissue deposit new matrix and angiogenesis provides for a renewed supply of nutrients and oxygen to the developing tissue. The ingrowth and activities of both fibroblasts and endothelial cells are apparently dependent upon the release of signals from activated platelets and macrophages[94,95] and triggered by changes in oxygen tension[96] or by substances released by damaged cells.[97] In addition, the elaboration of new matrix components influences ingrowth.[80,94,98] Eventually, a mature collagenous matrix is formed and redundant vessels are resorbed, leading to avascular scar tissue.

Of particular interest in these processes is angiogenesis, the formation of new blood vessels in the developing tissue. Angiogenesis is known to be induced by several of the growth and migratory factors released during inflammation,[94,95] as well as by matrix changes.[94] The ingrowth of new capillaries into granulation tissue requires an increase in growth and migration in the normally quiescent endothelium. The potential involvement of junctional transfer in the control of capillary endothelial growth state has been recently reported by Orlidge and D'Amore.[99] They have shown, in culture, that microvascular pericytes inhibit capillary endothelial cell growth by a contact-mediated phenomenon. Apparently, contact between these two cells results in elaboration of activated TGF-β,[100] a potent inhibitor of endothelial cell growth. While it is not known whether this contact-mediated release requires junctional communication, we have shown that microvascular endothelial cells and pericytes are capable of reciprocal dye transfer *in vitro*[71,75] and *in vivo*.[69,71,72] One possible brief scenario follows these lines: (1) wounding causes dissociation of the endothelial-pericyte contacts, resulting in release of endothelial quiescence (due to decreased levels of TGF-β); (2) local matrix turnover, increased in the wound area,[101] releases fibroblast growth factor (FGF) bound to basement membrane heparins,[102,103] also leading to increased endothelial growth; and (3) during maturation of the new capillaries, pericytes reappear around the endothelial cells, leading to inhibition of growth. The latter events are also aided by cessation of angiogenic stimulation (from platelets and macrophages as oxygen tension returns to normal) and the elaboration of mature matrix (which binds FGF) as the vessels stabilize.[94,95] While the involvement of junctional communication in these processes is by no means assured, the contact-dependent modulation of endothelial cell growth by pericytes is highly suggestive.

4. Regeneration

This process is essentially the replacement of lost and damaged tissue and cells. In wounds involving an epithelium, reconstitution of the barrier function of this tissue begins soon after the initial insult. The degree of damage, and hence the speed of regeneration, can determine the extent of some of the later stages in healing. If only the epithelium is damaged, these "erosions" heal directly and rapidly by regeneration: proliferation and migration of surrounding epithelial cells. In deeper wounds, the extent of regeneration depends on the extent of injury. In wounds of primary intention (e.g., by scalpel), the clean cut edges can result in rapid reconstitution of the epithelium and minimal scar tissue formation. In secondary intention wounds with ragged edges, or where missing tissue needs to be replaced, the slower reconstitution not only puts the organism at greater risk of infection, but also often results in increased scar tissue formation. To a degree, re-epithelialization is a classic *in vivo* experiment in growth control,[104] since a relatively stable condition is disrupted, leading to cell growth and migration, leading to reconstitution of the layer, leading to arrest of growth.

Several studies have suggested correlations between the size or frequency of gap junctions and the status of regeneration in a wide variety of tissues and cell sheets. Unfortunately, a simple correlation does not arise since both increases and decreases in the extent of gap junctions have been reported. Similarly, studies assessing coupling in regenerating systems have reported increases, decreases, or no change in junctional transfer. Essentially, two theses have been proposed. The first is that if gap junctional transfer is a negative modulator of growth (and hence involved in contact inhibition of growth), gap junctions and junctional transfer should be lost or diminished in actively mitotic, growing systems.[22-25] The second thesis, in analogy with that proposed for the interaction of epithelial cells or myofibroblasts in wound contraction, is that the regeneration of a wound requires coordinated action of the migrating, proliferating cells, suggesting that junctional transfer should be retained or even increased.[41,43,93,105] While we cannot as yet resolve this question, it may be that both camps are correct, and that the answer lies in the quantitative dynamics of the junctions, rather than in qualitative losses or increases. In addition, the variety of other factors, including cell adhesion molecules, growth and migratory factors, and matrix, which can modify the regeneration response, means that the effects of junctions, if any, are felt in concert rather than as a solo performance.

a. Skin and Other Epithelia

In contrast to the situation in hepatocytes and endothelium described below, ultrastructural studies on regenerating epithelia have found generally increased gap junctional extent.[2,41,43,93,105] These increases have been postulated to be important in control of coordinated migration[41,105] and mitotic activity.[43] Gabbiani and co-workers[41] reported a nearly fourfold increase in the proportion of epithelial cell surface occupied by gap junctions during wound regeneration in mammalian skin. Andersen[105] also reported an increase in gap junctions in migrating palatal epithelium. In a fascinating study on regeneration in rat tracheal epithelium, Gordon and co-workers[43] described the neoformation of gap junctions, in response to a wound, between cells that normally have essentially no gap junctions. Moreover, since their wounded tracheal epithelial cells proliferated synchronously through one wave of DNA synthesis, the authors were able to demonstrate formation of gap junctions by the end of S-phase and their disappearance during M-phase. If restimulated to go through another round, junctions began to appear during G_1 and reached a peak at the end of the second S. Subsequent differentiation of the cells into the normal pseudostratified epithelium resulted in the loss of gap junctions. A similar loss of gap junctions during differentiation has been reported for other epithelial cells,[37,106,107] but the correlation between the appearance of gap junctions and specific stages of the cell cycle strongly suggests a role of junctional transfer in the control of mitotic activity in this system.[43]

b. Hepatocytes

The regenerating liver is an interesting model for reconstitution of organ mass and organization. Partial hepatectomy causes a burst of mitotic activity for rapid but controlled growth.[109,110] Yee and Revel[47] showed that during this regenerative response in weanling rats, there was a clear and predictable disappearance and reappearance of gap junctions between adjacent hepatocytes. Further studies[48-50] demonstrated that these gap junctions disappeared between 24 and 28 h posthepatectomy and reappeared between 36 and 44 h. Studies on the extent of junctional transfer in regenerating liver have indicated either no change[51] or a decrease in electrotonic coupling[49] or dye transfer.[49]

Using quantitative Western blots, Traub et al.[110] demonstrated loss and reappearance of gap junction protein over a time course similar to that from the ultrastructural studies. The correlation between the expression of junction protein and the mitotic activity was clarified by Dermietzel et al.[111] who showed, both in regenerating liver and in cultured hepatocytes, that hepatocytes that are mitotically active (labeling with antibodies to bromodeoxyuridine incorporated into DNA) show a significant reduction in immunofluorescently detectable gap junction protein. Cells which were not mitotically active showed normal levels and distribution of gap junction protein. These correlative studies, and others linking alterations in gap junctions with other types of wounds or damage to hepatocytes,[112,113] provide interesting pointers to a direct involvement of gap junctions and junctional transfer in control of mitosis and regeneration in this system.

Studies on cultured hepatocytes have been useful in dissecting some of the relationships. Normally, when hepatocytes are placed into primary culture, the expression of differentiated characteristics is quite labile; gap junctions are lost within hours[114,115] as are other characteristics (it is known that gap junction proteins have a short (5 h) half-life in normal liver[116] so these structures are apparently normally quite labile). Interestingly, treatment of cultured hepatocytes with chondroitin sulfate proteoglycan, dermatan sulfate proteoglycan, or liver-derived heparin[114,115] resulted in nearly undiminished junctional transfer and clear stimulation of synthesis and expression of junction protein (Northern and Western blots, respectively). Thus certain extracellular matrix components modulate junctional transfer and other differentiated characteristics in these liver cells. It may be that, during regeneration of the liver, the loss of gap junctions may be related to the wound event, mitotic activity, and to the necessity to migrate over provisional matrix. The reestablishment of junctions may be related to the reestablishment of a normal matrix and a normal quiescent mitotic state.

c. Endothelium

Various studies have suggested that loss of gap junctions is an early event in the regenerative response of endothelium to intimal denudation,[117-119] paralleling the findings in regenerating liver mentioned above. In freeze-fracture electron microscopic studies on wounded endothelium in the rabbit common carotid, Spagnoli et al.[117] showed a clear decrease in the density and surface area of gap junctions, together with a loss of tight junctions, between cells at the migrating edge of the endothelium at 2 d after wounding. At 7 d, they found numerous small endothelial gap junctions and by 15 d, when restitution of endothelial continuity was complete, both gap and tight junctions were indistinguishable from controls. In general, the endothelial mitotic index was inversely related to the extent of gap junctions in these studies. The clear implication from these studies was that well-developed gap junctions inhibited endothelial cell growth. However, Hüttner and co-workers,[120] also using freeze-fracture electron microscopy, reported that the average width (as an index of size) of gap junctions increased in rat thoracic aortic endothelium at 15 d postwounding, compared to controls. This latter study did not involve a time course. The differences between these two results have not been resolved although they may relate to the differences in the morphometric techniques, species, and vessels.

Two subsequent studies have approached this question from the standpoint of junctional transfer rather than junctional size.[30,121] Jackman et al.[121] studied regeneration of endothelium in explanted rat aortas. They found undiminished capability for nucleotide transfer in cells near the wound edge as compared to controls, implying unimpaired junctional transfer. These results parallel early work by Stoker[122] using nucleotide transfer in wounded cultures of 3T3 cells. Studies in my laboratory have suggested that the capability for junctional transfer of microinjected Lucifer yellow is slightly but significantly diminished in wounded bovine aortic endothelial cultures. We found a 12% decrease in permeable interfaces between cells, over 2 d, within the first 20 rows from the wound edge.[30] In row 2, the decrease was 16%, and for cells that had separated from the migrating sheet, the decrease was 21%. Nevertheless, endothelial cells seem to maintain patent junctional contact in this stressed condition even while actively migrating to fill the wound, suggesting that junctional transfer may have a role in coordination of repair. Analysis of the data from this latter study has suggested to us that the decrease in dye transfer may be due to an increased turnover of gap junctions as the cells translocate past one another. Studies to test this hypothesis using cDNA probes for connexin43 mRNA (the junctional message made by endothelium *in vivo* and *in vitro*)[77] are underway. At this point, it is impossible to say whether the changes in the extent of junctions, or the extent of junctional transfer, are causally related to the changes in cell behavior or the obverse.

Recently, Shivers et al.[123] reported an influence of matrix components and soluble factors on elaboration of gap and tight junctions in cultured endothelium. In this study, large vessel endothelial cells, grown in astrocyte-conditioned medium and on an endothelial-derived extracellular matrix preparation, developed large complex tight junctions and large gap junctions. In their hands, cells grown on plastic or fibronectin had no junctions at all. These findings may be significant for wound repair *in vivo* (cf. the hepatocyte data above), since endothelium normally rests on its own matrix but when regenerating, migrates over provisional, fibronectin-rich matrix.

Of course, endothelial cell growth and migration are known to be modulated by a variety of soluble growth and migration factors, and membrane and matrix molecules,[124] so the experimental resolution of the control of these processes is not simple.

Junctional transfer may be only one of many, contributing control mechanisms. To further complicate matters, endothelial cells in small and large vessels apparently make gap junctions with subendothelial smooth muscle cells and pericytes.[62,63,125-129] Transfer through these heterocellular junctions has been shown in a few cases.[69-72,130,131] Therefore, it seems possible (as mentioned above in the example of capillary endothelial/pericyte combinations) that endothelial growth state is normally partially modulated by junctional transfer with the underlying cells; this control mechanism would of course also be disrupted in wounding and regeneration.

IV. CONCLUSIONS AND FUTURE DIRECTIONS

In reading through a recent volume on wound repair,[1] I was struck by the incredible variety of intercellular communication required for orderly restitution of damaged tissue. Soluble mediators from damaged cells, plasma constituents, inflammatory cells, and stromal cells all carry messages influencing repair. Similarly, the elaboration of provisional or mature matrix by one cell type communicates something to other cells that have to migrate over or attach to that substrate. Amidst this flurry of communication, the role(s) of junctional communication remain somewhat obscure. While we can demonstrate that gap junctions are induced or repressed in certain cells during some of the events described above, and presumably the functions of junctional transfer are augmented or decreased, we still do not have the crucial missing link between the existence or extent of junctions and transfer and the roles of transfer in communication. When junctions and transfer decrease during regeneration in the liver, is it because such a reduction is necessary and sufficient for release of growth quiescence or because the cells are too actively migrating to bother with making junctions? If junctions are necessary for coordination of regeneration, then why do they disappear during liver regeneration?

These questions will not be answered easily. One approach might be the use of specific junctional blockers. However, current modifiers of junctional permeability (such as tumor promoters, aliphatic alcohols, carbon tetrachloride, and agents that increase intracellular H^+, Ca^{2+}, or cAMP) all are confoundingly nonspecific. Injection of anti-junction antibodies[132] is specific but impractical for large-scale experiments. Pending the availability of such a usable mechanism (perhaps the specific deletion of genes), we will have to continue to rely on largely correlative studies. Nevertheless, as mentioned above, the response of tissues to wounding provides a complex but natural classroom for the study of intercellular communication and has and will continue to provide interesting tests for the roles of junctional communication.

REFERENCES

1. **Clark, R. A. F. and Henson, P. M., Eds.,** *The Molecular and Cellular Biology of Wound Repair,* Plenum Press, New York, 1988.
2. **McNutt, N. S. and Weinstein, R. S.,** Membrane ultrastructure at mammalian intercellular junctions, *Prog. Biophys. Molec. Biol.,* 26, 45, 1973.
3. **Larsen, W. J.,** Biological implications of gap junction structure, distribution and composition. A review, *Tissue Cell,* 15, 645, 1983.
4. **Zampighi, G. A. and Simon, S. A.,** The structure of gap junctions as revealed by electron microscopy, in *Gap Junctions,* Bennett, M. V. L. and Spray, D. C., Eds., Cold Spring Harbor Laboratory, Cold Spring Harbor, NY, 1985, 13.
5. **Caspar, D. L. D., Sosinsky, G. E., Tibbitts, T. T., Phillips, W. C., and Goodenough, D. A.,** Gap junction structure, in *Gap Junctions,* Hertzberg, E. L. and Johnson, R. G., Eds., Alan R. Liss, New York, 1988, 117.
6. **Flagg-Newton, J., Simpson, I., and Loewenstein, W. R.,** Permeability of the cell-to-cell membrane channels in mammalian cell junction, *Science,* 205, 404, 1979.
7. **Socolar, S. J. and Loewenstein, W. R.,** Methods for studying transmission through permeable cell-to-cell junctions, *Meth. Membr. Biol.,* 10, 123, 1979.
8. **Llinás, R. R.,** Electrotonic transmission in the mammalian central nervous system, in *Gap Junctions,* Bennett, M. V. L. and Spray, D. C., Eds., Cold Spring Harbor Laboratory, Cold Spring Harbor, NY, 1985, 337.
9. **Bennett, M. V. L., Zimering, M. B., Spira, M. E., and Spray, D. C.,** Interaction of electical and chemical synapses, in *Gap Junctions,* Bennett, M. V. L. and Spray, D. C., Eds., Cold Spring Harbor Laboratory, Cold Spring Harbor, NY, 1985, 355.
10. **Pitts, J. D.,** Direct interaction between animal cells, in *Cell Interactions,* Silvestri, L. G., Ed., Elsevier/North-Holland, Amsterdam, 1972, 227.
11. **Subak-Sharpe, J. H., Burk, R. R., and Pitts, J. D.,** Metabolic co-operation by cell to cell transfer between genetically different mammalian cells in tissue culture, *Heredity,* 21, 342.
12. **Ledbetter, M. L. and Lubin, M.,** Transfer of potassium. A new measure of cell-cell coupling, *J. Cell Biol.,* 80, 150, 1979.
13. **Lawrence, T. S., Beers, W. H., and Gilula, N. B.,** Transmission of hormonal stimulation by cell-to-cell communication, *Nature,* 272, 501, 1978.
14. **Murray, S. A. and Fletcher, W. H.,** Hormone-induced intercellular signal transfer dissociates cyclic AMP-dependent protein kinase, *J. Cell Biol.,* 98, 1710, 1984.
15. **Barr, L., Dewey, M. M., and Berger, W.,** Propagation of action potentials and the structure of the nexus in cardiac muscle, *J. Gen. Physiol.,* 48, 797, 1965.
16. **DeHaan, R. L.,** Dynamic behavior of cardiac gap junction channels, in *Gap Junctions,* Hertzberg, E. L. and Johnson, R. G., Eds., Alan R. Liss, New York, 1988, 305.
17. **Cole, W. C. and Garfield, R. E.,** Alterations in coupling in uterine smooth muscle, in *Gap Junctions,* Bennett, M. V. L. and Spray, D. C., Eds., Cold Spring Harbor Laboratory, Cold Spring Harbor, NY, 1985, 215.
18. **Neild, T. O.,** The relation between the structure and innervation of small arteries and arterioles and the smooth muscle membrane potential changes expected at different levels of sympathetic nerve activity, *Proc. R. Soc. London Ser. B,* 220, 237, 1983.
19. **Tomita, T.,** Electrical properties of mammalian smooth muscle, in *Smooth Muscle,* Bulbring, E., Brading, A. F., Jones, A. W., and Tomita, T., Eds., Edward Arnold, London, 1970, 197.
20. **Meda, P., Perrelet, A., and Orci, L.,** Endocrine cell interaction in the islets of Langerhans, in *The Functional Integration of Cells in Animal Tissues,* Pitts, J. D. and Finbow, M. E., Eds., Cambridge University Press, Cambridge, 1982, 113.
21. **Meda, P., Bruzzone, R., Chanson, M., and Bosco, D.,** Junctional coupling and secretion of pancreatic acinar cells, in *Gap Junctions,* Hertzberg, E. L. and Johnson, R. G., Eds., Alan R. Liss, New York, 1988, 353.
22. **Loewenstein, W. R.,** Junctional intercellular communication and the control of growth, *Biochim. Biophys. Acta,* 560, 1, 1979.
23. **Sheridan, J. D.,** Cell coupling and cell communication during embryogenesis, in *Cell Surface in Animal Embryogenesis and Development,* Poste, G. and Nicholson, G. L., Eds., North-Holland, Amsterdam, 1976, 409.

24. **Pitts, J. D., Kam, E., and Morgan, D.**, The role of junctional communication in cellular growth control and tumorigenesis, in *Gap Junctions*, Hertzberg, E. L. and Johnson, R. G., Eds., Alan R. Liss, New York, 1988, 353.

25. **Sheridan, J. D.**, Cell communication and growth, in *Cell-to-Cell Communication*, De Mello, W. C., Ed., Plenum Press, New York, 1987, 187.

26. **Loewenstein, W. R.**, Permeability of membrane junctions, *Ann. N.Y. Acad. Sci.*, 137, 441, 1966.

27. **De Mello, W. C.**, Cell-to-cell communication in heart and other tissues, *Prog. Biophys. Molec. Biol.*, 39, 147, 1982.

28. **Imanaga, I.**, Cell-to-cell diffusion of Procion yellow in sheep and calf Purkinje fibers, *J. Membr. Biol.*, 16, 381, 1974.

29. **El-Fouly, M. H., Trosko, J. E., and Chang, C.-C.**, Scrape-loading and dye transfer. A rapid and simple technique to study gap junctional intercellular communication, *Exp. Cell Res.*, 168, 422, 1987.

30. **Larson, D. M. and Haudenschild, C. C.**, Junctional transfer in wounded cultures of bovine aortic endothelial cells, *Lab. Invest.*, 59, 373, 1988.

31. **Haudenschild, C. C. and Harris-Hooker, S.**, Injury affects endothelial integrity beyond physical location of the damage, *Circulation*, 66II, 205, 1982.

32. **Haudenschild, C. C. and Harris-Hooker, S.**, Endothelial cell motility, in *The Biology of Endothelial Cells*, Jaffe, E. A., Ed., Martinus Nijhoff, The Hague, 1984, 74.

33. **Ryan, U. S., Absher, M., Olazabal, B. M., Brown, L. M., and Ryan, J. W.**, Proliferation of pulmonary endothelial cells: time-lapse cinematography of growth to confluence and restitution of monolayer after wounding, *Tissue Cell*, 14, 637, 1982.

34. **Stenn, K. S. and Depalma, L.**, Re-epithelialization, in *The Molecular and Cellular Biology of Wound Repair*, Clark, R. A. F. and Henson, P. M., Eds., Plenum Press, New York, 1988, 321.

35. **Hollenberg, N. K. and Odori, T.**, Centripetal spread of arterial collateral endothelial cell hyperplasia after renal artery stenosis in the rat, *Circ. Res.*, 60, 398, 1987.

36. **Paskins-Hurlburt, A. J., Hollenberg, N. K., and Abrams, H. L.**, Centripetal spread of endothelial cell mitotic activity in the artery leading to a rapidly growing tumor, *Microvasc. Res.*, 28, 131, 1984.

37. **Atkinson, M. M.**, Calcium fluxes induced in contiguous NRK cells following impalement of a single cell with a micropipette, *J. Cell Biol.*, 107, 792a, 1988.

38. **White, F. H., Thompson, D. A., and Gohari, K.**, Ultrastructural morphometry of gap junctions during differentiation of stratified squamous epithelium, *J. Cell Sci.*, 69, 67, 1984.

39. **Shimono, M. and Clementi, F.**, Intercellular junctions in oral epithelium. I. Studies with freeze fracture and tracing methods of normal rat keratinised oral epithelium, *J. Ultrastruct. Res.*, 56, 121, 1976.

40. **Caputo, R. and Peluchetti, D.**, The junctions of normal human epidermis: a freeze fracture study, *J. Ultrastruct. Res.*, 61, 44, 1977.

41. **Gabbiani, G., Chaponnier, C., and Hüttner, I.**, Cytoplasmic filaments and gap junctions in epithelial cells and myofibroblasts during wound healing, *J. Cell Biol.*, 76, 561, 1978.

42. **Elias, P. M. and Friend, D. S.**, The permeability barrier and pathways of transport in mammalian epidermis, *J. Cell Biol.*, 65, 180, 1975.

43. **Gordon, R. E., Lane, B. P., and Marin, M.**, Regeneration of rat tracheal epithelium: changes in gap junctions during specific phases of the cell cycle, *Exp. Lung Res.*, 3, 47, 1982.

44. **Ito, S. and Loewenstein, W. R.**, Ionic communication between early embryonic cells, *Devel. Biol.*, 19, 228, 1969.

45. **Kam, E., Melville, L., and Pitts, J. D.**, Patterns of junctional communication in skin, *J. Invest. Dermatol.*, 87, 748, 1986.

46. **Kam, E., Watt, F. M., and Pitts, J. D.**, Patterns of junctional communication in skin: studies on cultured keratinocytes, *Exp. Cell Res.*, 173, 431, 1987.

47. **Yee, A. G. and Revel, J. P.**, Loss and reappearance of gap junctions in regenerating liver, *J. Cell Biol.*, 78, 554, 1978.

48. **Yancey, S. B., Nicholson, B. J., and Revel, J.-P.**, The dynamic state of liver gap junctions, *J. Supramolec. Struct. Cell. Biochem.*, 16, 221, 1981.

49. **Meyer, D. J., Yancey, S. B., and Revel, J.-P.**, Intercellular communication in normal and regenerating rat liver: a quantitative analysis, *J. Cell Biol.*, 91, 505, 1981.

50. **Yancey, S. B., Easter, D., and Revel, J.-P.**, Cytological changes in gap junctions during liver regeneration, *J. Ultrastruct. Res.*, 67, 229, 1979.

51. **Loewenstein, W. R. and Penn, R. D.,** Intercellular communication and tissue growth. II. Tissue regeneration, *J. Cell Biol.,* 33, 235, 1967.

52. **Graf, J. and Petersen, O. H.,** Cell membrane potential and resistance in liver, *J. Physiol. (London),* 284, 105, 1978.

53. **Simionescu, N. and Simionescu, M.,** The cardiovascular system, in *Histology. Cell and Tissue Biology,* 5th ed., Weiss, L., Ed., Elsevier, New York, 1983, 371.

54. **Simionescu, M., Simionescu, N., and Palade, G. E.,** Segmental differentiations of cell junctions in the vascular endothelium. The microvasculature, *J. Cell Biol.,* 67, 863, 1975.

55. **Simionescu, M., Simionescu, N., and Palade, G. E.,** Segmental differentiations of cell junctions in the vascular endothelium. Arteries and veins, *J. Cell Biol.,* 68, 705, 1976.

56. **Schneeberger, E. E.,** Segmental differentiations of endothelial intercellular junctions in intra-acinar arteries and veins of the rat lung, *Circ. Res.,* 49, 1102, 1981.

57. **Hüttner, I., Boutet, M., and More, R. H.,** Gap junctions in arterial endothelium, *J. Cell Biol.,* 57, 247, 1973.

58. **Schneeberger, E. E. and Karnovsky, M. J.,** Substructure of intercellular junctions in freeze-fractured alveolar capillary membranes of mouse lung, *Circ. Res.,* 38, 404, 1976.

59. **Dermietzel, R.,** Junctions in the central nervous system of the cat. IV. Interendothelial junctions of cerebral blood vessels from selected areas of the brain, *Cell Tiss. Res.,* 164, 45, 1975.

60. **Hüttner, I., Boutet, M., and More, R. H.,** Studies on protein passage through arterial endothelium. I. Structural correlates of permeability in rat arterial endothelium, *Lab. Invest.,* 28, 672, 1973.

61. **Mink, D., Schiller, A., Kriz, W., and Taugner, R.,** Interendothelial junctions in kidney vessels, *Cell Tiss. Res.,* 236, 567, 1984.

62. **Rhodin, J. A. G.,** The ultrastructure of mammalian arterioles and precapillary sphincters, *J. Ultrastruct. Res.,* 18, 181, 1967.

63. **Rhodin, J. A. G.,** Ultrastructure of mammalian venous capillaries, venules and small collecting veins, *J. Ultrastruct. Res.,* 25, 452, 1968.

64. **Tani, E., Yamagata, S., and Ito, Y.,** Freeze-fracture of capillary endothelium in rat brain, *Cell Tiss. Res.,* 176, 157, 1977.

65. **Freddo, T. F. and Raviola, G.,** Freeze-fracture analysis of the interendothelial junctions in the blood vessels of the iris in *Macaca mulatta, Invest. Ophthalmol. Vis. Sci.,* 23, 154, 1982.

66. **Heinrich, D., Metz, J., Raviola, E., and Forssmann, W. G.,** Ultrastructure of perfusion-fixed fetal capillaries in the human placenta, *Cell Tiss. Res.,* 172, 157, 1976.

67. **Firth, J. A., Bauman, K. F., and Sibley, C. P.,** The intercellular junctions of guinea-pig placental capillaries: a possible structural basis for endothelial solute permeability, *J. Ultrastruct. Res.,* 85, 45, 1983.

68. **Raviola, G. and Raviola, E.,** Paracellular route of aqueous outflow in the trabecular network and canal of Schlemm. A freeze-fracture study of the endothelial junctions in the sclerocorneal angle of the macaque monkey eye, *Invest. Ophthalmol. Vis. Sci.,* 21, 52, 1981.

69. **Sheridan, J. D. and Larson, D. M.,** Junctional communication in the peripheral vasculature, in *The Functional Integration of Cells in Animal Tissues,* Pitts, J. D. and Finbow, M. E., Eds., Cambridge University Press, Cambridge, 1982, 263.

70. **Larson, D. M.,** Intercellular junctions and junctional transfer in the blood vessel wall, in *Endothelial Cells,* Vol. 3, Ryan, U. S., Ed., CRC Press, Boca Raton, FL, 1988, 75.

71. **Larson, D. M., Carson, M. P., and Haudenschild, C. C.,** Gap junctions in endothelial cells and pericytes, in *Microcirculation — An Update,* Vol. 1, Tsuchiya, M., Asano, M., Mishima, Y., and Oda, M., Eds., Elsevier Science Publishers, Amsterdam, 1987, 101.

72. **Larson, D. M. and Haudenschild, C. C.,** Junctional communication in the vessel wall in situ, *J. Cell Biol.,* 105, 327a, 1987.

73. **Larson, D. M. and Sheridan, J. D.,** Intercellular junctions and transfer of small molecules in primary vascular endothelial cultures, *J. Cell Biol.,* 92, 183, 1982.

74. **Larson, D. M. and Sheridan, J. D.,** Junctional transfer in cultured vascular endothelium. II. Dye and nucleotide transfer, *J. Membr. Biol.,* 83, 157, 1985.

75. **Larson, D. M., Carson, M. P., and Haudenschild, C. C.,** Junctional transfer of small molecules in cultured bovine brain microvascular endothelial cells and pericytes, *Microvasc. Res.,* 34, 184, 1986.

76. **Larson, D. M., Kam, E. Y., and Sheridan, J. D.,** Junctional transfer in cultured vascular endothelium. I. Electrical coupling, *J. Membr. Biol.,* 74, 103, 1983.

77. **Larson, D. M., Haudenschild, C. C., and Beyer, E. C.,** Vessel wall cells express mRNA for the connexin43 gap junction protein, *J. Cell Biol.,* 107, 556a, 1988.

78. **Terkeltaub, R. A. and Ginsberg, M. H.,** Platelets and response to injury, in *The Molecular and Cellular Biology of Wound Repair,* Clark, R. A. F. and Henson, P. M., Eds., Plenum Press, New York, 1988, 35.

79. **Williams, T. J.,** Factors that affect vessel reactivity and leukocyte emigration, in *The Molecular and Cellular Biology of Wound Repair,* Clark, R. A. F. and Henson, P. M., Eds., Plenum Press, New York, 1988, 115.

80. **Riches, D. W. H.,** The multiple roles of macrophages in wound healing, in *The Molecular and Cellular Biology of Wound Repair,* Clark, R. A. F. and Henson, P. M., Eds., Plenum Press, New York, 1988, 213.

81. **Kawahara, Y., Kariya, K., Araki, S., Fukuzaki, H., and Takai, Y.,** Platelet-derived growth factor (PDGF)-induced phospholipase C-mediated hydrolysis of phosphoinositides in vascular smooth muscle cells. Different sensitivity of PDGF- and angiotensin II- induced phospholipase C reactions to protein kinase C-activating phorbol esters, *Biochem. Biophys. Res. Commun.,* 156, 846, 1988.

82. **Campbell, F. R.,** Intercellular contacts between migrating blood cells and cells of the sinusoidal wall of bone marrow. An ultrastructural study using tannic acid, *Anat. Rec.,* 203, 365, 1982.

83. **Campbell, F. R.,** Intercellular contacts of lymphocytes during migration across high-endothelial venules of lymph nodes. An electron microscopic study, *Anat. Rec.,* 207, 643, 1983.

84. **Guinan, E. C., Smith, B. R., Davies, P. F., and Pober, J. S.,** Cytoplasmic transfer between endothelium and lymphocytes: quantitation by flow cytometry, *Am. J. Pathol.,* 132, 406, 1988.

85. **Skalli, O. and Gabbiani, G.** The biology of the myofibroblast relationship to wound contraction and fibrocontractive diseases, in *The Molecular and Cellular Biology of Wound Repair,* Clark, R. A. F. and Henson, P. M., Eds., Plenum Press, New York, 1988, 373.

86. **Crocker, D. J., Murad, T. M., and Geer, J. C.,** Role of the pericyte in wound healing. An ultrastructural study, *Exp. Mol. Pathol.,* 13, 51, 1970.

87. **Baur, P. S., Parks, D. H., and Hudson, J. D.,** Epithelial mediated wound contraction in experimental wounds — the purse-string effect, *J. Trauma,* 24,713, 1984.

88. **Ryan, G. B., Cliff, W. J., Gabbiani, G., Irle, C., Montandon, D., Statkov, P. R., and Majno, G.,** Myofibroblasts in human granulation tissue, *Hum. Pathol.,* 5, 55, 1974.

89. **Gabbiani, G., Hirschel, B. J., Ryan, G. B., Statkov, P. R., and Majno, G.,** Granulation tissue as a contractile organ. A study of structure and function, *J. Exp. Med.,* 135, 719, 1972.

90. **Gabbiani, G. and Rungger-Brändle, E.,** The fibroblast, in *Handbook of Inflammation: Tissue Repair and Regeneration,* Vol. 3, Glynn, L. E., Ed., Elsevier/North Holland, Amsterdam, 1981, 1.

91. **Bell, E., Ivarsson, B., and Merrill, C.,** Production of a tissue-like structure by contraction of collagen lattices by human fibroblasts of different proliferative potential in vivo, *Proc. Natl. Acad. Sci. U.S.A,* 76, 1274, 1979.

92. **Bellows, C. G., Melcher, A. H., Bhargava, U., and Aubin, J. E.,** Fibroblasts contracting three-dimensional collagen gels exhibit ultrastructure consistent with either contraction or protein secretion, *J. Ultrastruct. Res.,* 78, 178, 1982.

93. **Mahrle, G.,** Demonstration of giant and anular nexus in psoriatic epidermis, *Arch. Derm. Res.,* 261, 181, 1978.

94. **Madri, J. A. and Pratt, B. M.,** Angiogenesis, in *The Molecular and Cellular Biology of Wound Repair,* Clark, R. A. F. and Henson, P. M., Eds., Plenum Press, New York, 1988, 337.

95. **Zetter, B. R.,** Angiogenesis. State of the art, *Chest,* 93, 159s, 1988.

96. **Knighton, D. R., Hunt, T. K., Scheuenstuhl, H., Halliday, B. J., Werb, Z., and Banda, M. J.,** Oxygen tension regulates the expression of angiogenesis factor by macrophages, *Science,* 221, 1283, 1983.

97. **D'Amore, P. A. and Thompson, R. W.,** Collateralization in peripheral vascular disease, in *Vascular Diseases: Current Research and Clinical Applications,* Strandness, D. E., Ed., Grune & Stratton, New York, 1987, 319.

98. **McCarthy, J. B., Sas, D. F., and Furcht, L. T.,** Mechanisms of parenchymal cell migration into wounds, in *The Molecular and Cellular Biology of Wound Repair,* Clark, R. A. F. and Henson, P. M., Eds., Plenum Press, New York, 1988, 281.

99. **Orlidge, A. and D'Amore, P. A.,** Inhibition of capillary endothelial cell growth by pericytes and smooth muscle cells, *J. Cell Biol.,* 105, 1455, 1987.

100. **Orlidge, A., Smith, S., and D'Amore, P. A.,** Co-cultures of endothelial cells and pericytes produce activated TGF-B, *J. Cell Biol.,* 107, 504a, 1988.

101. **Oosta, G. M., Favreau, L. V., Beeler, D. L., and Rosenberg, R. D.,** Purification and properties of human platelet heparitinase, *J. Biol. Chem.,* 257, 11249, 1982.

102. **Vlodavsky, I., Folkman, J., Sullivan, R., Fridman, R., Ishai-Michaeli, R., Sasse, J., and Klagsbrun, M.,** Endothelial cell-derived basic fibroblast growth factor synthesis and deposition into subendothelial extracellular matrix, *Proc. Natl. Acad. Sci. U.S.A,* 84, 2292, 1987.

103. **Baird, A. and Ling, N.,** Fibroblast growth factors are present in the extracellular matrix produced by endothelial cells in vitro: implications for a role of heparinase-like enzymes in the neovascular response, *Biochem. Biophys. Res. Commun.,* 142, 428, 1987.

104. **Todaro, G. J., Lazar, G. K., and Green, H.,** The initiation of cell division in a contact-inhibited mammalian cell line, *J. Cell. Comp. Physiol.,* 66, 325, 1965.

105. **Andersen, L.,** Cell junctions in squamous epithelium during wound healing in palatal mucosa of guinea pigs, *Scand. J. Dent. Res.,* 88, 328, 1980.

106. **Morgan, G., Pearson, J. D., and Sepulveda, F. V.,** Absence of gap junctional complexes in two established renal epithelial cell lines (LLC-PK1 and MDCK), *Cell Biol. Int. Rep.,* 8, 917, 1984.

107. **Cereijido, M., Robbins, E., Sabatini, D. D., and Stefani, E.,** Cell-to-cell communication in monolayers of epithelioid cells (MDCK) as a function of the age of the monolayer, *J. Membr. Biol.,* 81, 41, 1984.

108. **Bucher, N. L.,** Experimental aspects of hepatic regeneration, *N. Engl. J. Med.,* 277, 686, 1967.

109. **Bucher, N. L.,** Experimental aspects of hepatic regeneration, *N. Engl. J. Med.,* 277, 738, 1967.

110. **Traub, O., Drüge, P. M., and Willecke, K.,** Degradation and resynthesis of gap junction protein in plasma membrane of regenerating liver after partial hepatectomy or cholestasis, *Proc. Natl. Acad. Sci. U.S.A,* 80, 755, 1983.

111. **Dermietzel, R., Yancey, S. B., Traub, O., Willecke, K., and Revel, J.-P.,** Major loss of the 28-kD protein of gap junction in proliferating hepatocytes, *J. Cell Biol.,* 105, 1925, 1987.

112. **Metz, J. and Bressler, D.,** Reformation of gap and tight junctions in regenerating liver after cholestasis, *Cell Tiss. Res.,* 199, 257, 1979.

113. **Schellens, J. P. M., Blangé, T., and de Groot, K.,** Gap junction ultrastructure in rat liver parenchymal cells after in vivo ischemia, *Virchows Arch. B,* 53, 347, 1987.

114. **Spray, D. C., Fujita, M., Saez, J. C., Choi, H., Watanabe, T., Hertzberg, E., Rosenberg, L. C., and Reid, L. M.,** Proteoglycans and glycosaminoglycans induce gap junction synthesis and function in primary liver cultures, *J. Cell Biol.,* 105, 541, 1987.

115. **Fujita, M., Spray, D. C., Choi, H., Saez, J. C., Watanabe, T., Rosenberg, L. C., Hertzberg, E. L., and Reid, L. M.,** Glycosaminoglycans and proteoglycans induce gap junction expression and restore transcription of tissue-specific mRNAs in primary liver cultures, *Hepatology,* 7, 1S, 1987.

116. **Fallon, R. G. and Goodenough, D. A.,** Five-hour half-life of mouse liver gap-junction protein, *J. Cell Biol.,* 90, 521, 1981.

117. **Spagnoli, L. G., Pietra, G. G., Villaschi, S., and Johns, L. W.,** Morphometric analysis of gap junctions in regenerating arterial endothelium, *Lab. Invest.,* 46, 139, 1982.

118. **Schwartz, S. M., Stemerman, M. B., and Benditt, E. P.,** The aortic intima. II. Repair of the aortic lining after mechanical denudation, *Am. J. Pathol.,* 81, 15, 1975.

119. **Schwartz, S. M., Haudenschild, C. C., and Eddy, E. M.,** Endothelial regeneration. I. Quantitative analysis of initial stages of endothelial regeneration in rat aortic intima, *Lab. Invest.,* 38, 568, 1978.

120. **Hüttner, I., Walker, C., and Gabbiani, G.,** Aortic endothelial cell during regeneration. Remodeling of cell junctions, stress fibers, and stress fiber-membrane attachment domains, *Lab. Invest.,* 53, 287, 1985.

121. **Jackman, R W., Anderson, S. K., and Sheridan, J. D.,** The aortic intima in organ culture. Response to culture conditions and partial endothelial denudation, *Am. J. Pathol.,* 133, 241, 1988.

122. **Stoker, M. P.,** The effects of topoinhibition and cytochalasin B on metabolic cooperation, *Cell,* 6, 253, 1975.

123. **Shivers, R. R., Arthur, F. E., and Bowman, P. D.,** Induction of gap junctions and brain endothelium-like tight junctions in cultured bovine endothelial cells: local control of cell specialization, *J. Submicrosc. Cytol. Pathol.,* 20, 1, 1988.

124. **Heimark, R. L. and Schwartz, S. L.,** The role of cell-cell interaction in the regulation of endothelial cell growth, in *The Molecular and Cellular Biology of Wound Repair,* Clark, R. A. F. and Henson, P. M., Eds., Plenum Press, New York, 1988, 359.

125. **Spagnoli, L. G., Villaschi, S., Neri, L., and Palmieri, G.,** Gap junctions in myo-endothelial bridges of rabbit carotid arteries, *Experientia,* 38, 124, 1982.

126. **Taugner, R., Kirchheim, H., and Forssmann, W.,G.,** Myoendothelial contacts in glomerular arterioles and in renal interlobular arteries of rat, mouse and Tupaia belangeri, *Cell Tiss. Res.,* 235, 319, 1984.

127. **Spitznas, M. and Reale, E.,** Fracture faces of fenestrations and junctions of endothelial cells in human choroidal vessels, *Invest. Ophthalmol.,* 14, 98, 1975.

128. **Tilton, R. G., Kilo, C., and Williamson, J. R.,** Pericyte-endothelial relationships in cardiac and skeletal muscle capillaries, *Microvasc. Res.,* 18, 325, 1979.

129. **Cuevas, P., Gutierrez-Diaz, J. A., Reimers, D., Dujovny, M., Diaz, F. G., and Ausman, J. I.,** Pericyte endothelial gap junctions in human cerebral capillaries, *Anat. Embryol.,* 170, 155, 1984.

130. **Davies, P. F., Olesen, S. P., Clapham, D. E., Morrel, E. M., and Schoen, F. J.,** Endothelial communication. State of the art lecture, *Hypertension,* 11, 563, 1988.

131. **Davies, P. F.,** Vascular cell interactions with special reference to the pathogenesis of atherosclerosis, *Lab. Invest.,* 55, 5, 1986.

132. **Warner, A. E., Guthrie, S. C., and Gilula, N. B.,** Antibodies to gap junctional protein selectively disrupt communication in the early amphibian embryo, *Nature,* 311, 127, 1984.

Chapter 7

CHEMICAL, ONCOGENE AND GROWTH REGULATOR MODULATION OF EXTRACELLULAR, INTRACELLULAR AND INTERCELLULAR COMMUNICATION

James E. Trosko, Chia-Cheng Chang, Burra V. Madhukaar, and Saw Yin Oh

TABLE OF CONTENTS

"...what mysterious forces precede the appearance of the processes, promote their growth and ramification, stimulate the corresponding migration of the cells and fibers in predetermined directions, as if in obedience to a skillfully arranged architectural plan, and finally establish those protoplasmic kisses, the intercellular articulations, which seem to constitute the final ecstasy of an epic love story?"

Santiago Ramon Y. Cajal[1]

I. INTRODUCTION: HIERARCHICAL HOMEOSTASIS

While one of the major challenges in biology is to understand the complex regulatory processes within a cell, one can easily see that the problem is amplified by orders of magnitude when one tries to understand how a multicellular organism orchestrates the regulatory processes between like and unlike cells. The philosophical/ scientific concepts to help explain this complex problem of regulation of ions and molecules at the molecular/biochemical levels, at the cellular levels, at the organ and systems level, and at the whole organism/environment levels have been the hierarchical and cybernetic principles.[2-6] This is further illustrated by the insight provided by H. Eagle[7] when he stated: "It is a truism that in biology the whole is greater than the sum of its parts. Just as the exploration of molecular behavior cannot predict the precisely ordered interaction between cellular organelles, so the study of discrete cells will not suffice to explain the interaction of many different cell types complexly associated into organs which in turn exercise a mutual control."

Scientifically, at the molecular/biochemical level, various cybernetic feedback systems have been described to explain how gene expression can be regulated by the consequence of the expression or how the products of an enzyme's activity regulate the enzyme. At the cellular level, proliferation of normal stem cells in either solid or soft tissue seems to be under negative control by either direct contact ("contact inhibition")[8,9] or indirect suppression by negative growth regulators from their differentiated progeny.[10-16] At the organ or system level, cells of one tissue type can control the growth or differentiation of distal cells of another tissue type by either positive or negative growth regulators or effectors (i.e., hormones, peptide regulators). All of these systems within the organism, of course, are subject to environmental intervention by physical and chemical agents which can act as negative or positive influences at each of the biological feedback systems (Figure 1).

With this as a brief background, this chapter explicates a hypothesis which tries to integrate three known regulatory processes of multicellular organisms, namely (1) the *extracellular* communication processes, involving molecular signals such as hormones, peptide growth regulators, neurotransmitters, etc.; (2) *intracellular* communication mechanisms, involving various second messengers such as intracellular free Ca^{++}, cyclic AMP, inositol triphosphate, etc.; and (3) the *intercellular* communication process, involving the passage of ions and small molecular weight molecules via the membrane-associated protein structure, the gap junction[17] (Figure 2). In short, while others have offered integrating hypotheses, none have involved the role of gap

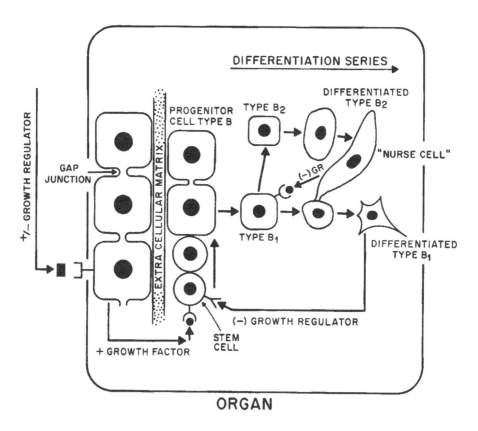

FIGURE 1. A diagram to illustrate the interrelationship of intercellular communication by positive and negative growth regulators on stem cell growth and differentiation and intercellular communication via gap junctions in a closed organ system. (From Trosko, J. E. and Chang, C. C., in *Biologically Based Methods for Cancer Risk Assessment,* Travis, C., Ed., Lewis Publishers, Chelsea, MI, 1989. With permission.).

junctional intercellular communication.[18-21] This hypothesis could provide an explanation for integrating multiple systems, such as the hypothalamic-pituitary-gonadal-thymic systems and the neuroendocrine-immune systems and other multisystem functions, as well as explain how the disruption of any one system can affect all dependent and interactive systems. It simply implies that extracellular factors, endogenous or exogenous, by triggering various transmembrane signaling mechanisms, can modulate gap junction function which, in turn, regulates a variety of cellular functions.

II. ROLE OF EXTRACELLULAR, INTRACELLULAR, AND INTERCELLULAR COMMUNICATION IN GROWTH CONTROL AND DIFFERENTIATION

A. EXTRACELLULAR COMMUNICATION

Claude Bernard first introduced the idea that there had to be a cybernetic-like control system in multicellular organisms in order to regulate, adaptively, growth, development, maintenance of normal function, and wound healing.[2] The term *homeostasis* was

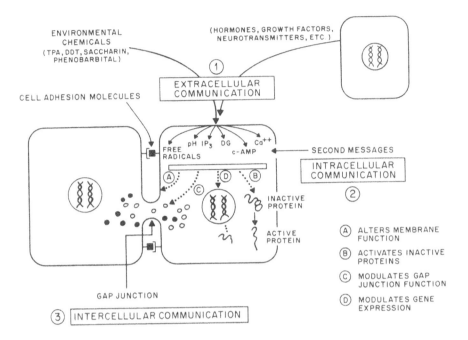

FIGURE 2. The heuristic schemata characterizes the postulated link between extracellular communication and intercellular communication via various intracellular transmembrane signaling mechanisms. It provides an integrating view of how the neuroendocrine-immune system ("mind or brain/body connection") and other multi-system coordinations could occur. Although not shown here, activation or altered expression of various oncogenes (and "anti-oncogenes") could also contribute to the regulation of gap junction function.

formulated by W.B. Cannon[3] to characterize this regulatory process. Weiss and Kavanau[10] provided, conceptually, growth regulators of cell proliferation, and Mazia predicted that the stimulation of cell division in a multicellular organism was the result of the "removal of a block" (see Reference 23). The conceptualization of positive and negative factors, which were produced in cells of one tissue which affected distal cells of another tissue, to explain how a constant "milieu interieur" is maintained predated the discovery of many of the actual molecular entities involved in homeostasis. With the discovery of various hormones, biologically active peptides (neuropeptides, growth factors), lymphokines, and neurotransmitters, it was possible to flesh out the concept of homeostasis by cybernetic processes with specific molecular details. It then became possible to explain how the secretion of specific molecules by one cell which specifically interacted with another target cell triggered a physiological response, such as stimulation or inhibition of growth, induction of differentiation, or the production of secretory products. The affected cells, in turn, secrete other products which affect the original signaling cell, thereby completing the cybernetic loop. The mechanisms by which each of these extracellular communicating signals (i.e., hormones, growth factors, neurotransmitters, etc.) elicit their physiological responses are usually mediated by specific membrane or cytoplasmic receptor molecules. The hormone, growth factor or neurotransmitter-receptor complex triggers a variety of intracellular biochemical mechanisms.

B. INTRACELLULAR COMMUNICATION

After exposure of a cell to either endogenous or exogenous chemicals, the cell has a variety of means to adapt to environmental changes. The most obvious is the interaction of extracellular signaling molecules to specific membrane receptors. Three basic mechanisms have been characterized by which the transmembrane receptor complex generates the intracellular signal: (1) ligand-modulated ion channel activity, (2) ligand-regulated enzymatic activity, and (3) ligand-regulated with other submembraneous constituents like the G-components of the adenylate cyclase system.[24]

Once the ligand-receptor complex has been formed, a number of "second messengers" has been identified which seem to be the signal for triggering a new physiological state: (1) sodium ion, (2) cyclic AMP, (3) cyclic GMP, (4) calcium ion, (5) diacylglycerol, (6) inositol triphosphate, (7) a-substituents of the G-protein complex, and (8) other as yet unidentified peptide regulators related to the action of insulin. These messengers, in turn, affect a wide variety of enzymes or binding proteins, such as cyclic AMP dependent protein kinases, protein kinase C, calcium binding proteins, calmodulin, etc.

C. INTERCELLULAR COMMUNICATION

In multicellular organisms, not only is orchestration of cells between different tissues necessary (extracellular communication) but also the coordination of activities, proliferation and differentiation of cells within excitable and nonexcitable tissues. In fact, evidence is mounting to support the notion that the cellular consequence of extracellular communication, mediated by triggering intracellular communication, is the modulation of intercellular communication (Figure 3). The fact that various hormones,[25] neurotransmitters,[26] and peptide growth regulators have been shown to modulate intercellular communication[27] suggests that many more extracellular communicating signals (e.g., lymphokines, neuropeptides, chalones, etc.) probably modulate intercellular communication.

As defined in this report, intercellular communication is mediated by the membranous specialized protein structure, the gap junction. Cells coupled by this membrane channel, which permits passive diffusion of ions and small molecular weight molecules (<1200 D), are electronically and metabolically communicating.[28,29] This ability to share electrical and metabolic signals within a group of coupled cells allows for the separation of specialized functions within a multicellular organism, since the multicellular organism has developed means to prevent all cells from being gap junctionally coupled to each other. Physical extracellular matrices, as well as specific cell adhension molecules (CAMs),[30-32] would prevent two heterologous cell types from communicating via gap junctions. On the other hand, various heterologous cell types, by being coupled, can influence each other's development (e.g., granulosa-ovum cells[33]). The physiological consequence of having a group of cells coupled by gap junctions could be either to provide the means by which a ionic or molecular signal could emanate from a "source" to stimulate activity or to act as a "drain" or "sink" to dampen a signal to prevent some physiological response.[34]

While progress on the molecular characterization of the structure, protein, and gene(s) regulating the gap junction is being made,[35] definitive understanding of some fundamental questions still remain, such as "Is there but one gap junction protein (gene) which is not only conserved throughout evolution, but also found throughout all tissues within a species?"

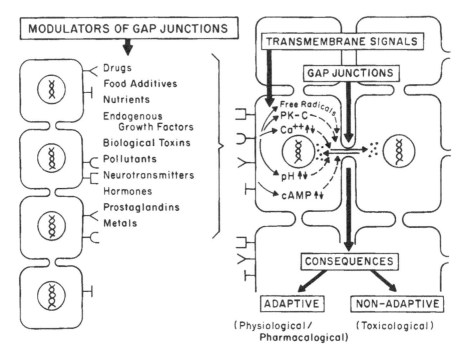

FIGURE 3. This diagram summarizes how a variety of chemicals might modulate gap junction function by receptor- or nonreceptor-mediated second messengers or by other mechanisms. The consequences of either decreasing or inducing gap junctional communication will depend on the circumstances (e.g., compensatory hyperplasia leading either to wound healing or to tumor promotion). (From Trosko, J. E., et al., in *Biochemical Mechanisms and Regulation of Intercellular Communication,* Milman, H. A. and Elmore, E., Eds., Princeton Scientific Publishers, Princeton, NJ, 1987. With permission.)

From the standpoint of the presumed role the gap junction plays in the regulation of cell growth and differentiation,[28,29,33,36-39] it would seem that it would be highly conserved, with regard to those regions of the protein needed for the placement in the membrane and extracellular coupling to the subunit (the connexin), yet having room for variation in that part of the protein subject to tissue specific regulation in the cytoplasm.

The biological role of gap junctions has been postulated to be closely linked to the regulation of cell growth, to the differentiation of cells, and to the synchronization of activities of cells.[36,40,41] Most of the evidence suggesting that gap junctions are involved in these fundamental biological functions of differentiated multicellular organisms can be considered indirect; some recent evidence, using antibodies to gap junction proteins and molecular biological techniques with cloned genes, suggest these roles are associated with the functional gap junction.[42,43]

The general observation that normal cells, which have functional gap junctions, can "contact-inhibit" growth,[44] while those with dysfunctional or decreased number of gap junctions do not contact-inhibit (e.g., cancer cells),[45-47] suggests that there is, at least, a correlation between growth control and the presence of gap junctions. In addition, cancer cells, which are cells unable to terminally differentiate, usually have a reduced number or dysfunctional gap junctions.[48] Antibodies directed against a gap junctional protein have been shown to also alter development and differentiation in several organisms.[42,43] More recently, Field,[49] using a fusion gene constructed of the transcriptional regulatory sequence of the atrial natriuretic factor and the SV40 T antigen (an oncoprotein), produced both tumors in the atrium and cardiac arrhythmias in a transgenic mouse transfected with this fusion gene. This observation is important in light of the fact that Azarnia and Loewenstein[50] have shown that the T antigen (large and small) could alter junctional permeability. Since it would be predicted that normal gap junctional communication would be expected for both growth control and atrial conduction, specific expression of a gene product which blocks gap junctions only in the heart tissue would be predicted to contribute to both of these diverse disease states.

In summary, as pointed out by T. Dobzhansky that "nothing in biology makes sense except in the light of evolution," the widespread association of gap junctions in most cells of all metazoans[31] and in the fundamental functions of these cells would seem to support the idea of its fundamental importance in evolution and adaptability of metazoans.

III. CANCER AS A DISEASE OF GROWTH CONTROL AND DIFFERENTIATION

Several basic observations related to carcinogenesis and concepts derived from these observations have led to a possible link between intercellular communication and cancer. The composite picture of carcinogenesis is that of a complex multistep process, involving the clonal expansion and evolution of a single stem-like cell. Epidemiological, experimental carcinogenesis studies, as well as *in vitro* transformation experiments, have suggested multiple gene[51-55] and/or epigenetic[56] events needed to complete the carcinogenic process.

The clonal nature of cells within a tumor has been determined in a number of cancers.[52,57] Heterogeneity within the tumor during clonal expansion is evidence of the evolution of genotypes and phenotypes within cells of the tumor.[58,59] This is probably due to the selection of those phenotypes best adapted to the changing microenvironment caused by the growth of the tumor.[59]

The stem cell theory of cancer is based on many divergent observations. Concepts such as "cancer as a disease of differentiation",[60,61] or as "oncogeny as partially blocked ontogeny",[62] implies that not all cells of the organism (pluripotent stem cells, progenitor cells, differentiated cells) are at equal risk for transformation. The observation that cancer cells within a given tissue type manifest various stages of differentiation (with the exception of the N-stage of that cell lineage) suggests that the carcinogenic process has induced a block in the cell's ability to terminally differentiate, but not to self-renew (Figure 4).

The fact that many tumor cells, *in vitro*, have been induced by a variety of agents to terminally differentiate[63-67] indicates they possess the genetic capabilities to terminally differentiate, yet *in vivo* are unable to respond to those endogenous factors triggering differentiation.

One of the other phenotypes of a cancer cell which is relevant to this analysis is the inability of the cancer cell to be contact-inhibited. In effect, while a normal cell senses the presence of a normal neighboring cell and ceases to proliferate, a cancer cell appears insensitive to the physical (and chemical) contact with the surrounding normal cells.

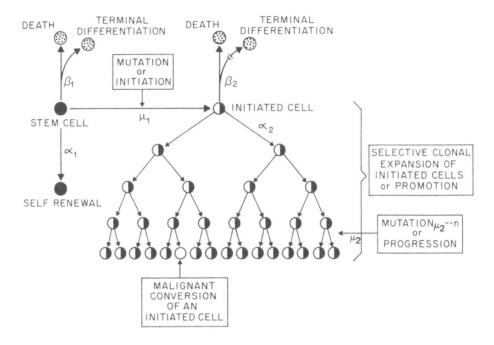

FIGURE 4. The initiation/promotion/progression model of carcinogenesis. β_1 = rate of terminal differentiation and death of stem cell; β_2 = rate of death, but not of terminal differentiation of the initiated cell $(-//\rightarrow)$; α_1 = rate of cell division of stem cells; α_2 = rate of cell division of initiated cells; μ_1 = rate of the molecular event leading to initiation (i.e., possibly mutation); μ_2 = rate at which second event occurs within an initiated cell. (From Trosko, et al., in *Gap Junctions*, Hertzberg, E. L. and Johnson, R. G., Eds., Alan R. Liss, New York, 1988. With permission.)

Correlated with this loss of growth control (and inability to terminally differentiate) has been the demonstration of abnormal gap junctional intercellular communication of tumor cells.[48] The dysfunctional intercellular communication could either be the result of (1) the presence of a factor which, selectively, down-regulates gap junctional communication in the tumor cell, (2) a genetic defect in the tumor cell which prevents a tumor cell from communicating with a normal cell (i.e., selective communication due to a cell adhension molecule defect[68,69]), (3) a genetic defect in either the gap junction gene or some regulation of the gene, or (4) the genetic deficiency to respond to molecular signals transmitted via gap junctions.

To summarize this section, the cancer-associated phenotypes of the loss of growth control and of the inability to terminally differentiate seems to correlate with the abnormal gap junctional communication found in these cells.

IV. THE MULTISTEP PROCESS OF CARCINOGENESIS, CHEMICAL TUMOR PROMOTERS, AND INTERCELLULAR COMMUNICATION

The concepts of initiation/promotion/progression have been operationally defined from observations on experimental animals exposed to a variety of carcinogenic factors.[70-72] Although at present the underlying molecular mechanisms for each of these phases of carcinogenesis are not known, they clearly are very different. Assuming the initiating event takes place in a single stem or progenitor cell the experimental observations in mouse skin suggests that the event is ostensibly *irreversible*. The fact that most, if not all, true mutagens are good initiators[54,73] and "complete" carcinogens (when given at high enough concentrations to induce cell death),[74] suggests that *mutagenesis* is the underlying mechanism for initiation.[75]

Promotion has been defined as a potentially interruptible or reversible process which enhances the frequency and earlier appearance of tumors in the initiated animal.[70-73,76,77] One hypothesis is that promotion allows for the clonal expansion of the initiated stem cell either by selective accumulation of this cell which can self-renew but not terminally differentiate[73,78] or by prevention of apoptosis (programmed cell death).[79] The papilloma in the skin, the enzyme-altered foci in the liver, and the polyps of the colon might be viewed as the tissue manifestation of a single stem cell which, when stimulated to proliferate or prevented from dying, accumulates in the tissue. Promotion is the result, among other possible processes, primarily a sustained mitogenic stimulus.[78,80]

Lastly, progression is that event which converts a promoter-dependent, premalignant, initiated cell to a malignant and tumor-promoter independent cell. Mutagens, rather than promoters, appear to be most efficient in converting premalignant cells to carcinomas.[81] This is also evidence that at least two genetic events are needed for the completion of carcinogenesis.

A major new insight to the linkage of gap junctional communication and carcinogenesis came when, independently, Yotti et al.[82] and Murray and Fitzgerald[83] observed that the mouse skin tumor promoter, 12-tetradecanoylphorbol-13-acetate (TPA), could inhibit gap junctional intercellular communication *in vitro*. This observation has been validated in other cells using different techniques to measure gap junction function (e.g., metabolic cooperation,[84-95] electrocoupling,[96] dye injection,[97] scrape-loading/dye transfer,[98] and fluorescence recovery after photobleaching[99]) (Figure 5). Later, it was shown that a wide variety of chemicals, such as natural plant and animal toxins, drugs, dietary chemicals, food additives, pollutants, solvents, metabolites, hormones, growth factors, neurotransmitters, and heavy metals, have been shown to inhibit gap junctional intercellular communication.[95] While not all of these chemicals are known to be tumor promoters, most are known to be physiological modulators or toxicants of a wide sort (e.g., teratogens, immune modulators, neuro- and reproductive toxicants). It is important to note that most known tumor promoters, which work at noncytotoxic levels *in vivo*, are inhibitors of intercellular communication. Furthermore, it must be emphasized that any procedure which causes cell proliferation *in vivo* (cell death brought about by a wide variety of causes [genotoxicants, burning, freezing, nonspecific cytotoxins and cytotoxicants, specific inhibitors of critical cellular functions], cell removal [surgery, wounding], etc.),[78] could act as a tumor-promoting stimulus, since in each of these cases, cell to cell communication is disrupted. All of these means to kill or remove cells are known to induce compensatory or regenerative hyperplasia in the surviving stem cell by allowing the cells to escape the contact inhibition of the surrounding and communicating cells.

The demonstration that the tumor-promoting agent TPA reduced the frequency of gap junctions *in vitro* in Chinese hamster cells[100] and in mouse skin *in vivo*[101] supports the hypothesis that promotion involves the inhibition of gap junctional communica-

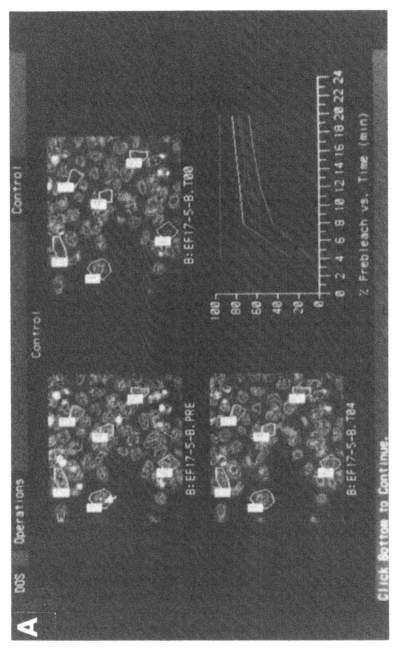

FIGURE 5. Fluorescence images demonstrating the use of the Meridian ACAS™ Interactive Laser Cytometer to measure gap junctional intercellular communication. Human epithelial kidney cells panel were fluorescently labeled with 6-carboxyfluorescein. The upper panel (A) shows recovery of fluorescence through gap junctions after photobleaching the dye in cells 2 through 6. These cells were then incubated for 1 h with 10 ng/ml of TPA. The lower panel (B) shows that TPA blocked the gap junction-mediated transfer of dye in these cells.

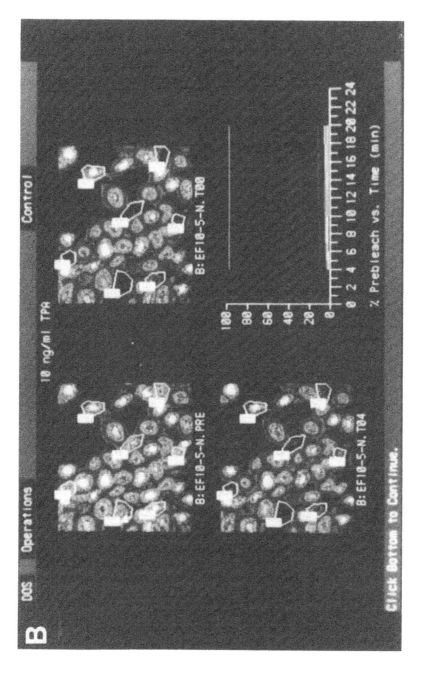

FIGURE 5 (continued)

tion. This has been confirmed by using monoclonal antibodies to gap junctions[102] and freeze-fracture analysis[103] to note that the number of gap junctions have been reduced in hepatocarcinomas in initiated/promoted rats. Furthermore, the amount of gap junction mRNA was reduced in primary tumors of initiated/promoted rat livers.[104]

In summary, elimination of contact inhibition, mediated by gap junctions, between an initiated cell and its surrounding normal neighbors, either by cell removal or cell death or by noncytotoxic means with exogenous chemicals, can isolate the initiated cell. This isolation could allow this cell to proliferate and selectively accumulate, since by definition it will not be removed from the population as would normal cells because of its inability to terminally differentiate.

V. ROLE OF ONCOGENES AND GROWTH FACTORS IN CARCINOGENESIS

The concept of oncogenes, i.e., sequences of DNA which have an operational association with cancer, has forced us to try to integrate the initiation/promotion/ progression concept with the genes which might be involved in the discrete phases of carcinogenesis.[105,106] Physical and chemical initiators seem to block the ability of stem cells to terminally differentiate.[54,73,107-112] Chemical promoters appear to block contact inhibition and allow cell growth. During progression, a genomic stabilization of this blockage or intercellular communication occurs, rendering the cell independent of external chemical promoters. Alteration of specific genes (oncogenes) could affect each of these functions. The oncogenes so far identified seem to be related to growth factors, growth factor receptors, signal transductor elements, and transcription and replication elements,[113] which is not surprising, since they have been isolated operationally in cells which have selective growth advantages, either by not terminally differentiating or by loss of growth control.

Since chemical tumor promoters have been shown to act either as growth or differentiation regulators, in part by their ability to modulate gap junctional intercellular communication,[78] then oncogenes, which also have been associated with either growth or differentiation of cells,[114-116] might be mediating their affects via modulation of gap junctional communication. The src, ras, neu, and mos oncogenes,[117,122] as well as the polyomavirus middle T gene,[123] have been shown in several *in vitro* cell systems to be acting like TPA, in that gap junctional intercellular communication is dysfunctional in all these systems.

The stable loss of gap junctional communication has been associated with tumor metastasis.[124-126] Since the reversible inhibition of cell communication has been associated with the tumor promotion phase, and since transformed cells *in vitro* seem to be suppressed by surrounding and communicating normal cells and show selective communication at the morphologically transformed stage,[97] one might postulate that the initiation phase might be related to those oncogenes which affect terminal differentiation. The fact is that in some systems the myc oncogene seems to be needed for "immortalization" whereas the ras is needed for transformation.[127,128] In other systems, the ras oncogene alone seems to be able, in an apparent one-step process, to initiate or transform cells.[129,130] Two major problems arise with these observations. First, does the carcinogenesis process first involve the immortalization of a normal mortal cell or is it the prevention of the mortalization of a normal immortal stem cell?

Second, when ras acts as a "complete" carcinogen, is it because the cell in which it acts already has a high level of endogenous myc-like activity or because it has no antisuppressor or antioncogene expressed?[131,132] Until these questions are resolved, little can be done to determine which stage the various oncogenes affect.

Since several oncogenes, which code for growth factors, growth factor receptors, or growth-related biochemical functions, seem to modulate gap junction function, it would seem reasonable to predict that the growth factors, the cellular means for extracellular communication, would be able to modulate intercellular communication. By definition, growth factors are those cellular products which can induce a contact-inhibited cell to enter into the mitogenic phase. Therefore, assuming gap junctions are necessary to maintain contact inhibition, one would predict that growth factors would inhibit gap junction function. This prediction has been borne out in experiments using epidermal growth factor (EGF), bovine pituitary extracts, and transforming growth factor beta (TGF-β) on human keratinocytes in serum-free medium.[27] Under these conditions, EGF and bovine pituitary extract acted as stimulators of cell growth, while TGF-β acted to induce differentiation in these cells. Certain hormones, which can be considered growth factors for certain cells, have been known to modulate gap junctional communication.[133] Finally, similar to chemical tumor promoters and several oncogenes, growth factors and hormones have been shown to act as tumor promoters *in vitro*[134-136] and *in vivo*.[137-142]

VI. POSSIBLE MECHANISMS OF ACTION OF MODULATORS OF INTERCELLULAR COMMUNICATION: FUNCTIONS OF EXTRACELLULAR, INTRACELLULAR, AND INTERCELLULAR COMMUNICATION

As has been previously stated, homeostasis in a multicellular organism depends on cells of one tissue to communicate, with positive and negative signals (extracellular communication), to cells in other tissues (intercellular communication) via transmembrane signaling mechanisms (intracellular communication). These important multicellular functions are achieved via the coordination of these three communication mechanisms after the fertilization of the egg to the adaptive functions of a mature organism, namely, the control of proliferation of the single zygotic cell, the facilitation of areas of differentiation, and the adaptive control functions in the differentiated cells. As was indicated earlier, one important consequence of extracellular communication in the multicellular organism was the isolation of gap-junctionally coupled cells so that the extracellular triggered transmembrane signals could reach critical masses in order to physiologically change the target cell. Without the ability to downregulate gap junctions, the transmembrane signals would be dampened by diffusion to the other cells. While extracellular factors would induce either proliferation or differentiation by modulating intercellular communication (e.g., chemical tumor promoters, EGF or TGF-β), the intracellular signals triggered by specific extracellular messengers on specific cells would determine if the consequence would be proliferation or differentiation.

The inhibition of gap junctional intercellular communication via intracellular communicating signals would allow for intracellular regulatory signals to be increased. To affect this shift in the physiological equilibrium, the extracellular communication signals must (1) bring about an influx of ions and substrate molecules for mitogenesis or differentiation, (2) prevent the dilution of the intracellular regulatory ions and molecules by blocking gap junction transfer, (3) activate inactive proteins and enzymes, and (4) modulate gene expression.[143] Coupled with these ideas is the hypothesis by Binggeli and Weinstein[144] that the gap junction channel becomes the sodium channel when the cell is uncoupled. This would explain the relationship of these two functionally different types of membrane channels and the inhibition or stimulation of DNA synthesis.

At this point, our knowledge of the exact mechanisms by which gap junction is regulated is very limited. However, one fact seems clear: namely, there is more than one mechanism which can affect gap junction function. This seems not only clear from the wide variety of genetic or developmental factors, hormones, neurotransmitters, activated oncogenes, and growth factors, but also the wide variety of exogenous chemicals which have been shown to modulate gap junction function.

Some of the earliest studies on gap junction regulation have indicated that intracellular pH_i, Ca^{++}, or asymmetric voltages between cells could alter its functioning.[145] Therefore, exposure to chemicals which could alter these parameters of the cell could affect gap junctional intercellular communication. If the internal changes induced by these chemicals were sublethal, depending on the cell, the consequence might be for the cell to proliferate, differentiate, or perform an adaptive differentiated function.

Recently, the demonstration that TPA, a powerful inhibitor of gap junction function, appears to mediate its effects through the activation of protein kinase C suggests that one of the proteins phosphorylated by the Ca^{++}-dependent, phospholipid-sensitive protein kinase is the gap junction protein or one of its regulatory proteins. It has been demonstrated, in a cell-free system, that a gap junction protein can be phosphorylated.[146]

Equally interesting are the observations that chemicals which increase cAMP levels in the cell to either suppress growth or induce differentiation while increasing gap junctional communication, activate a cAMP-dependent protein kinase which also can phosphorylate the gap junction protein.[147,148] These two afore-mentioned observations suggest that one mode of regulating gap junctional communication is a coupled phosphorylation/dephosphorylation system. It also suggests, in cases where differential responses to downregulation effects of TPA (i.e., transient vs. long-term downregulation), that the response might depend on different dephosphorylation capacities of different cells.

In addition to these potential mechanisms, Saez et al.[149] showed that carbon tetrachloride inhibited intercellular communication, not by increase in intracellular free Ca^{++} or acidity or by changing the phosphorylated state of the main gap junction protein, but apparently by the oxidation of sulfhydryl groups needed for gap junction coupling.

Strong evidence that there exist multiple mechanisms comes from studies showing that coupled cells, refractory to TPA, could be uncoupled by exposure to chemicals which do not act through protein Kinase-C, such as 4,4′-dichlorodiphenyltrichloroethane.[150,151] Also, kinetics of inhibition of gap junction function and its recovery indicates that several alcohols[152] and glycyrrhetinic acid might be modulating gap junctions by yet another mechanism.[153]

The downregulation of gap junction function by various oncogene products which are distinctly different (src, ras, mos, neu) suggests, again, other mechanisms of action. The fact that cellular src has been associated with differentiation or differentiated functions in the brain tissue suggests a developmental mechanism to control gap junction function. The fact that many oncogenes code for protein kinases indicates that phosphorylation of different amino acids in the gap junction protein, together with the other target substrates for these kinases, would affect the cell's function.

The recent experimental demonstration of a phenomenon of tumor suppression by the fusion of a tumor cell with a normal cell or the injection of specific normal genes into a tumor cell has led to the concept of antioncogenes or tumor-suppressor genes.[131,132,154] If one of the consequences of an oncogene is to downregulate gap junction function to reduce contact inhibition, then, conceptually, the tumor-suppressor genes would be to upregulate the gap junction. Future experiments should be able to test this prediction.

Given that there seem to be multiple mechanisms by which gap junctional communication, a fundamental cellular process, can be regulated by endogenous factors, it should not be surprising that many exogenous chemicals could mimic or interfere with this process at many steps (i.e., the extracellular or intracellular communication levels). Depending on important circumstances (e.g., concentration, time of exposure, interaction with other chemicals, species, sex, developmental stage, and tissue differences), the modulation of intercellular communication could have either adaptive (i.e., wound healing, growth) or maladaptive consequences, (i.e., teratogenesis, tumor promotion, neurotoxicity).

Therefore, it seems from an evolutionary viewpoint, this important biological function was designed to be sensitive to a wide variety of endogenous and exogenous extracellular signals, which could regulate the gap junction function via a number of intracellular messengers.

ACKNOWLEDGMENTS

The authors wish to thank our laboratory assistants, Mrs. Beth Lockwood, Ms. Heather Rupp, Mr. William Paradee, and Guixiang Zhang for their dedicated assistance, as well as our secretary, Mrs. Darla Conley, for her expert word-processing skills. The manuscript was based on research supported by grants from the National Cancer Institute (CA-21104) and the U.S. Air Force Office of Scientific Research (AFOSR 86-0084) and a gift from R.J. Reynolds Company.

REFERENCES

1. **Cajal, S. R.,** *Memoirs of the American Philosophical Society,* The American Philosophical Society, Philadelphia, 1937, 373.
2. **Bernard, C.,** *Sur les Phenomenes de la Vie,* J.B. Balliere et Fils, Paris, 1878.
3. **Cannon, W. B.,** Organization for physiological homeostasis, *Physiol. Rev.,* 9, 399, 1929.
4. **Potter, V. R.,** Probabilistic aspects of the human cybernetic machine, *Perspect. Biol. Med.,* 17, 164, 1974.
5. **Brody, H.,** A systems view of man: implications for medicine, science and ethics, *Perspect. Biol. Med.,* 17, 71, 1973.
6. **Iverson, O. H.,** Cybernetic aspects of the cancer problem, in *Progress in Biocybernetics,* Vol. 2, Wiemer, N. and Schade, J. P., Eds., Elsevier, Amsterdam, 1965, 76.
7. **Eagle, H.,** Growth regulatory effects of cellular interaction, *Israel J. Med. Sci.,* 1, 1220, 1965.
8. **Levine, E. M., Becker, Y., Boone, C. W., and Eagle, H.,** Contact inhibition, macromolecular synthesis and polyribosomes in cultured human diploid fibroblasts, *Proc. Natl. Acad. Sci. U.S.A.,* 53, 350, 1965.
9. **Abercrombie, M.,** Contact inhibition and malignancy, *Nature,* 281, 259, 1979.
10. **Weiss, P. and Kavanau, J. L.,** A model of growth and growth control in mathematical terms. *J. Gen. Physiol.,* 41, 1, 1957.
11. **Osgood, E. E.,** A unified concept of the etiology of the leukemias, lymphomas, and cancer, *J. Natl. Cancer Inst.,* 18, 155, 1957.
12. **Bullough, W. S.,** The chalones, *Sci. J.,* 5, 71, 1969.
13. **Iverson, O. H.,** Cellular growth and cancer formation, *Br. J. Hosp. Med.,* 3, 105, 1970.
14. **Potter, V. R.,** The present status of the blocked ontogeny hypothesis of neoplasia: the thalassemia connection, *Oncodevel. Biol. Med.,* 2, 243, 1981.
15. **Iverson, O. H.,** The chalones, in *Handbook of Experimental Pharmacology,* Vol. 57, Baserga, R., Ed., Springer-Verlag, Berlin, 1981, chap. 17.
16. **Potter, V. R.,** Cancer as a problem in intercellular communication: regulation by growth inhibiting factors, *Prog. Nucleic Acid Res. Molecul. Biol.,* 29, 161, 1983.
17. **Trosko, J. E., Chang, C. C., and Madhukaar, B. V.,** Chemical and oncogene modulation of intercellular communication in tumor promotion, in *Biochemical Mechanisms and Regulation of Intercellular Communication,* Milman, H. A. and Elmore, E., Eds., Princeton Scientific Publishers, Princeton, NJ, 1987, 209.
18. **Yabrov, A.,** Adequate function of the cell: interactions between the needs of the cell and the needs of the organism, *Med. Hypoth.,* 6, 337, 1980.
19. **Roth, J., LeRoith, D., Shiloach, J., and Rubinovitz, C.,** Intercellular communication: an attempt at a unifying hypothesis, *Clin. Res.,* 31, 354, 1983.
20. **Blalock, J. E.,** The immune system as a sensory organ, *J. Immunol.,* 132, 1067, 1984.
21. **Grossman, C. J.,** Interactions between the gonad steroids and the immune system, *Science,* 227, 257, 1985.
22. **Mazia, D.,** Mitosis and the physiology of cell division, in *The Cell,* Vol. III, Brachet, J. and Mirsky, A. E., Eds., Academic Press, New York, 1961, chap. 2.
23. **Iverson, O. H.,** The chalones, in *Handbook of Experimental Pharmacology,* Vol. 57, Baserga, R., Ed., Springer-Verlag, Berlin, 1981, chap. 17.
24. **Hollenberg, M. D.,** Mechanisms of receptor-mediated transmembrane signalling, *Experientia,* 42, 718, 1986.
25. **Larsen, W. J. and Risinger, M. A.,** The dynamic life histories of intercellular membrane junctions, *Mod. Cell. Biol.,* 4, 151, 1985.
26. **Neyton, J. and Trautman, A.,** Acetylcholine modulation of the conductances of intercellular junctions between rat lacrimal cells, *J. Physiol.,* 377, 283, 1986.
27. **Madhukaar, B. V., Oh, S. Y., Chang, C. C., Wade, M., and Trosko, J. E.,** Altered regulation of intercellular communication by epidermal growth factor, transforming growth factor-b and peptide hormones in normal human keratinocytes, *Carcinogenesis,* 10, 13, 1989.
28. **Loewenstein, W. R.,** Junctional intercellular communication and the control of growth, *Biochim. Biophys. Acta,* 560, 1, 1979.
29. **Loewenstein, W. R.,** Junctional intercellular communication: the cell to cell membrane channel, *Physiol. Rev.,* 61, 829, 1981.

30. **Edelman, G. M.,** Cell adhension molecules, *Science,* 219, 450, 1983.
31. **Obrink, B.,** Epithelial cell adhesion molecules, *Exp. Cell Res.,* 163, 1, 1986.
32. **Rutishauser, U., Acheson, A., Hall, A. K., Mann, D. M., and Sunshine, J.,** The neural cell adhesion molecule (NCAM) as a regulator of cell-cell interactions, *Science,* 240, 53, 1988.
33. **Schultz, R. M.,** Roles of cell to cell communication in development, *Biol. Reprod.,* 32, 27, 1985.
34. **Sheridan, J. D.,** Cell communication and growth, in *Cell-to-Cell Communication,* DeMello, W. C., Ed., Plenum Press, New York, 1987, 187.
35. **Hertzberg, E. L. and Johnson, R. G., Eds.,** *Gap Junctions,* Vol. 7, Alan R. Liss, New York, 1988.
36. **Pitts, J. D. and Finbow, M. E.,** The gap junction, *J. Cell Sci.,* 4, 239, 1986.
37. **MacDonald, C.,** Gap junctions and cell-cell communication, *Essays Biochem.,* 21, 86, 1985.
38. **Neyton, J. and Trautman, A.,** Physiological modulation of gap junction permeability, *J. Exp. Biol.,* 124, 43, 1986.
39. **Bennett, M. V. L. and Spray, D. C.,** *Gap Junctions,* Cold Spring Harbor Laboratory, Cold Spring Harbor, NY, 1985.
40. **Finbow, M. E. and Yancey, S. B.,** The role of intercellular junctions, in *Biochemistry of Cellular Regulation,* Vol. 4, Clemens, M. J., Ed., CRC Press, Boca Raton, FL, 1981, 215.
41. **Meda, P., Bruzzone, R., Chanson, M., and Bosco, D.,** Junctional coupling and secretion of pancreatic acinar cells, in *Gap Junctions,* Vol. 7, Hertzberg, E. L. and Johnson, R. G., Eds., Alan R. Liss, New York, 1988, 353.
42. **Warner, A. E., Guthrie, S. C., and Gilula, N. B.,** Antibodies to gap junctional protein selectively disrupt communication in the early amphibian embryo, *Nature,* 311, 127, 1984.
43. **Fraser, S. E., Green, C. R., Bode, H. R., and Gilula, N. B.,** Selective disruption of gap junctional communication interferes with a patterning process in hydra, *Science,* 237, 49, 1987.
44. **Levine, E. M., Becker, Y., Boone, C. W., and Eagle, H.,** Contact inhibition, macromolecular synthesis, and polyribosomes in cultured human diploid fibroblasts, *Proc. Natl. Acad. Sci. U.S.A.,* 53, 350, 1965.
45. **Borek, C. and Sachs, L.,** The difference in contact inhibition of cell replication between normal cells and cells transformed by different carcinogens, *Proc. Natl. Acad. Sci. U.S.A.,* 56, 1705, 1966.
46. **Corsaro, C. M. and Migeon, B. R.,** Comparison of contact-mediated communication in normal and transformed human cells in culture, *Proc. Natl. Acad. Sci. U.S.A.,* 74, 4476, 1977.
47. **Abercrombie, M.,** Contact inhibition and malignancy, *Nature (London),* 281, 259, 1979.
48. **Kanno, Y.,** Modulation of cell communication and carcinogenesis, *Jpn. J. Physiol.,* 35, 693, 1985.
49. **Field, L. J.,** Atrial natriuretic factor-SV40 T antigen transgenes produce tumors and cardiac arrhythmias in mice, *Science,* 239, 1029, 1988.
50. **Azarnia, R. and Loewenstein, W. R.,** Intercellular communication and the control of growth. XII. Alteration of junction permeability by simian virus 40. Roles of the large and small T antigens, *J. Membr. Biol.,* 82, 213, 1984.
51. **Revel, J.-P.,** The oldest multicellular animal and its junctions, in *Gap Junctions,* Vol. 7, Hertzberg, E. L. and Johnson, R. G., Eds., Alan R. Liss, New York, 1988, 135.
52. **Nowell, P. C.,** The clonal evolution of tumor cell populations, *Science,* 198, 23, 1976.
53. **Cairns, J.,** The origin of human cancers, *Nature,* 289, 353, 1981.
54. **Potter, V. R.,** A new protocol and its rationale for the study of initiation and promotion of carcinogenesis in rat liver, *Carcinogenesis,* 2, 1375, 1981.
55. **Moolgavkar, S. H.,** Carcinogenesis modeling: from molecular biology to epidemiology, *Annu. Rev. Pub. Health,* 7, 151, 1986.
56. **Kerbel, R. S., Frost, P., Liteplo, R., Carlow, D. A., and Elliott, B. E.,** Possible epigenetic mechanisms of tumor progression, *J. Cell Physiol.,* 3, 87, 1984.
57. **Fialkow, P. J.,** Clonal origins of human tumors, *Biochim. Biophys. Acta,* 458, 384, 1976.
58. **Fidler, I. J. and Hart, I. R.,** The development of biological diversity and metastatic potential in malignant neoplasms, *Oncodevel. Biol. Med.,* 4, 161, 1982.
59. **Nicolson, G. L.,** Tumor cell instability, diversification, and progression to the metastatic phenotype: from oncogene to oncofetal expression, *Cancer Res.,* 47, 1473, 1987.
60. **Markert, C.,** Neoplasia: a disease of cell differentiation, *Cancer Res.,* 28, 1908, 1968.
61. **Pierce, G. B.,** Neoplasms, differentiation and mutations, *Am. J. Pathol.,* 77, 103, 1974.
62. **Potter, V. R.,** Phenotypic diversity in experimental hepatomas: the concept of partially blocked ontogeny, *Br. J. Cancer,* 38, 1, 1978.

63. **Schubert, D. and Jacob, F.,** 5-Bromodeoxyuridine-induced differentiation of a neuroblastoma, *Proc. Natl. Acad. Sci. U.S.A.*, 67, 247, 1970.

64. **Huberman, E. and Callaham, M. F.,** Induction of terminal differentiation in human promyelocytic leukemia cells by tumor-promoting agents, *Proc. Natl. Acad. Sci. U.S.A.*, 76, 1293, 1979.

65. **Yamamoto, Y., Tomida, M., and Hozumi, M.,** Production by mouse spleen cells of factors stimulating differentiation of mouse myeloid leukemia cells that differ from the colony-stimulating factor, *Cancer Res.*, 40, 4804, 1980.

66. **Till, J. E.,** Stem cells in differentiation and neoplasms, *J. Cell Physiol.*, 1, 3, 1982.

67. **Bloch, A.,** Induced cell differentiation in cancer therapy, *Cancer Treat. Rep.*, 68, 199, 1984.

68. **Kanno, Y., Sasaki, Y., Shiba, Y., Yoshida-Noro, C., and Takeichi, M.,** Monoclonal antibody ECCD-1 inhibits intercellular communication in teratocarcinoma PCC3 cells, *Exp. Cell Res.*, 152, 270, 1984.

69. **Enomoto, T. and Yamasaki, H.,** Lack of intercellular communication between chemically transformed and surrounding non-transformed BALB/c3T3 cells, *Cancer Res.*, 44, 5200, 1984.

70. **Boutwell, R. K.,** The function and mechanisms of promoters of carcinogenesis, *CRC Crit. Rev. Toxicol.*, 2, 419, 1974.

71. **Slaga, T. J.,** *Mechanisms of Tumor Promotion*, Vol. 1, CRC Press, Boca Raton, FL, 1983.

72. **Pitot, H. C., Goldsworthy, T., and Moran, S.,** The natural history of carcinogenesis: implications of experimental carcinogenesis in the genesis of human cancer, *J. Supramolec. Cell. Biochem.*, 17, 133, 1981.

73. **Trosko, J. E. and Chang, C. C.,** An integrative hypothesis linking cancer, diabetes and atherosclerosis: the role of mutations and epigenetic changes, *Med. Hypoth.*, 6, 455, 1980.

74. **Trosko, J. E., Jone, C., and Chang, C. C.,** The role of tumor promoters on phenotypic and alterations affecting intercellular communication and tumorigenesis, *Ann. N.Y. Acad. Sci.*, 407, 316, 1983.

75. **Trosko, J. E. and Chang, C. C.,** Implications for risk assessment of genotoxic and non-genotoxic mechanisms in carcinogenesis, in *Methods for Estimating Risk of Chemical Inquiry: Human and Non-Human Biota and Ecosystems*, Vouk, V. B., Butler, G. C., Hoel, D. G., and Peakall, D. B., Eds., John Wiley & Sons, Chichester, England, 1985, 181.

76. **Berenblum, I. and Armuth, V.,** Two independent aspects of tumor promotion, *Biochim. Biophys. Acta*, 651, 51, 1981.

77. **Pitot, H. C. and Sirica, A. E.,** The stages of initiation and promotion in hepatocarcinogenesis, *Biochim. Biophys. Acta*, 605, 191, 1980.

78. **Trosko, J. E., Chang, C. C., and Medcalf, A.,** Mechanisms of tumor promotion: potential role of intercellular communication, *Cancer Invest.*, 1, 511, 1983.

79. **Bursch, W., Lauer, B., Timmermann-Trosieuer, I., Barthel, G., Schupplerr, J., and Schulte-Hermann, R.,** Controlled death (apoptosis) of normal and putative preneoplastic cells in rat liver following withdrawal of tumor promoters, *Carcinogenesis*, 5, 453, 1984.

80. **Frei, J. V. and Stephens, P.,** The correlation of promotion of tumour growth and of induction of hyperplasia in epidermal two-stage carcinogenesis, *Br. J. Cancer*, 22, 83, 1968.

81. **Hennings, H., Shores, R., Wenk, M. L., Spangler, E. F., Tarone, R., and Yuspa, S. H.,** Malignant conversion of mouse skin tumors is increased by tumor initiators and unaffected by tumor promoters, *Nature*, 304, 67, 1983.

82. **Yotti, L. P., Chang, C. C., and Trosko, J. E.,** Elimination of metabolic cooperation in Chinese hamster cells by a tumor promoter, *Science*, 208, 1089, 1979.

83. **Murray, A. W. and Fitzgerald, D. J.,** Tumor promoters inhibit metabolic cooperation in co-cultures of epidermal and 3T3 cells, *Biochem. Biophys. Res. Commun.*, 91, 395, 1979.

84. **Williams, G. M., Telang, S., and Tong, C.,** Inhibition of intercellular communication between liver cells by the liver promoter, 1,1,1-trichloro-2,2-bis(p-chlorophenyl)ethane, *Cancer Lett.*, 11, 339, 1981.

85. **Umeda, M., Noda, K., and Ono, T.,** Inhibition of metabolic cooperation in Chinese hamster cells by various chemicals including tumor promoters, *Gann*, 71, 614, 1980.

86. **Mosser, D. D., and Bols, N. C.,** The effect of phorbols on metabolic cooperation between human fibroblasts, *Carcinogenesis*, 3, 1207, 1982.

87. **Malcolm, A. R., Mills, L. J., and McKenna, E. J.,** Inhibition of metabolic cooperation between Chinese hamster V79 cells by tumor promoters and other chemicals, *Ann. N.Y. Acad. Sci.*, 407, 488, 1983.

88. **Davidson, J. S., Baumgarten, I., and Harley, E. H.**, Use of a new citrulline incorporation assay to investigate inhibition of intercellular communication by 1,1,1-trichloro-2,2-bis(p-chlorophenyl)-ethane in human fibroblasts, *Cancer Res.*, 45, 515, 1985.

89. **Gupta, R. S., Singh, B., and Stetsko, D. K.**, Inhibition of metabolic cooperation by phorbol esters in a cell culture system based on adenosine kinase deficient mutants of V79 cells, *Carcinogenesis*, 6, 1359, 1985.

90. **Hartman, T. G. and Rosen, J. D.**, The effect of some experimental parameters on the inhibition of metabolic cooperation by phorbol myristate acetate, *Carcinogenesis*, 6, 1315, 1985.

91. **Elmore, E., Milman, H. A., and Wyatt, G. P.**, Application of the Chinese hamster V79 metabolic cooperation assay in toxicology, in *Biochemical Mechanisms and Regulation of Intercellular Communication*, Milman, H. A. and Elmore, E., Eds., Princeton Scientific Publishers, Princeton, NJ, 1987, 265.

92. **Kavanagh, T. J., Chang, C. C., and Trosko, J. E.**, Characterization of a human teratocarcinoma cell assay for inhibitors of metabolic cooperation, *Cancer Res.*, 46, 1359, 1986.

93. **Ruch, R. J., Klaunig, J. E., and Pereira, M. A.**, Inhibition of intercellular communication between mouse hepatocytes by tumor promoters, *Toxicol. Appl. Pharm.*, 87, 111, 1987.

94. **Jone, C., Trosko, J. E., and Chang, C. C.**, Characterization of a rat liver epithelial cell line to detect inhibitors of metabolic cooperation, *In Vitro Cell Devel. Biol.*, 23, 214, 1987.

95. **Trosko, J. E. and Chang, C. C.**, Nongenotoxic mechanisms in carcinogenesis: role of inhibited intercellular communication, in *Banbury Report: New Directions in the Qualitative and Quantitative Aspects of Carcinogenic Risk Assessment*, Battey, R., Ed., Cold Spring Harbor Laboratory, Cold Spring Harbor, NY, in press.

96. **Enomoto, T., Sasaki, Y., Shiba, Y., Kanno, Y., and Yamasaki, H.**, Tumor promoters cause a rapid and reversible inhibition of the formation and maintenance of electrical cell coupling in culture, *Proc. Natl. Acad. Sci. U.S.A.*, 78, 5628, 1981.

97. **Enomoto, T. and Yamasaki, H.**, Lack of intercellular communication between chemically transformed and surrounding non-transformed BALB/c 3T3 cells, *Cancer Res.*, 44, 5200, 1984.

98. **El-Fouly, M. H., Trosko, J. E., and Chang, C. C.**, Scrape-loading and dye transfer: a rapid and simple technique to study gap junctional intercellular communication, *Exp. Cell Res.*, 168, 422, 1987.

99. **Wade, M. H., Trosko, J. E., and Schindler, M.**, A fluorescence photobleaching assay of gap junction-mediated communication between human cells, *Science*, 232, 525, 1986.

100. **Yancey, S. B., Edens, J. E., Trosko, J. E., Chang, C. C., and Revel, J.-P.**, Decreased incidence of gap junctions between Chinese hamster V79 cells upon exposure to the tumor promoter, TPA, *Exp. Cell Res.*, 139, 329, 1982.

101. **Kalimi, G. H. and Sirsat, S. M.**, The relevance of gap junctions to stage 1 tumor promotion mouse epidermis, *Carcinogenesis*, 5, 167, 1984.

102. **Janssen-Timmen, U., Traub, O., Dermietzel, R., Rabes, H. M., and Willecke, K.**, Reduced number of gap junctions in rat hepatocarcinomas detected by monoclonal antibody, *Carcinogenesis*, 7, 1475, 1986.

103. **Sugie, S., Mori, H., and Takahashi, M.**, Effect of in vivo exposure to the liver tumor promoters, phenobarbital or DDT on the gap junctions of rat hepatocytes: A quantitative freeze-fracture analysis, *Carcinogenesis*, 8, 45, 1987.

104. **Beer, D. G., Neveu, M. J., Paul, D. L., Rapp, U. R., and Pitot, H. C.**, Expression of the c-raf protooncogene, δ-glutamyltranspeptidase, and gap junction protein in rat liver neoplasms, *Cancer Res.*, 48, 1610, 1988.

105. **Trosko, J. E., Jone, C., and Chang, C. C.**, Oncogenes, inhibited intercellular communication and tumor promotion, in *Cellular Interactions by Environmental Tumor Promoters*, Fujiki, H., Hecker, E., Moore, R. E., Sugimura, T., and Weinstein, I. B., Eds., Japan Science Society Press, Tokyo, 1984, 101.

106. **Trosko, J. E., Chang, C. C., Madhukar, B. V., Oh, S. Y., Bombick, D., and El-Fouly, M. H.**, Modulation of gap junction intercellular communication by tumor promoting chemicals, oncogenes and growth factors during carcinogenesis, in *Gap Junctions*, Hertzberg, E.L., and Johnson, R.G., Eds., Alan R. Liss, New York, 1988, 435.

107. **Yuspa, S. H. and Morgan, D. L.**, Mouse skin cells resistant to terminal differentiation associated with initiation of carcinogenesis, *Nature*, 293, 72, 1981.

108. **Yuspa, S. H., Kilkenny, A. E., Stanley, J., and Lichti, U.,** Keratinocytes blocked in phorbol ester response early stage of terminal differentiation by sarcoma viruses, *Nature,* 314, 459, 1985.

109. **Scott, R. E. and Maercklein, P. B.,** An initiator of carcinogenesis selectively and stably inhibits stem cell differentiation: a concept that initiation of carcinogenesis involves multiple phases, *Proc. Natl. Acad. Sci. U.S.A.,* 82, 2995, 1985.

110. **Quintanilla, M., Brown, K., Ramsden, M., and Balmain, A.,** Carcinogen-specific mutation and amplification of Ha-ras during mouse skin carcinogenesis, *Nature,* 332, 78, 1986.

111. **Kawamura, H., Strickland, J. E., and Yuspa, S. H.,** Association of resistance to terminal differentiation with initiation of carcinogenesis in adult mouse epidermal cells. *Cancer Res.,* 45, 2748, 1985.

112. **Miller, D. R., Viaje, A., Aldaz, C. M., Conti, C. J., and Slaga, T. J.,** Terminal differentiation-resistant epidermal cells in mice undergoing two-stage carcinogenesis, *Cancer Res.,* 47, 1935, 1987.

113. **Weinberg, R. A.,** The action of oncogenes in the cytoplasm and nucleus, *Science,* 230, 770, 1985.

114. **Ohlsson, R. I. and Pfeifer-Ohlsson, S. B.,** Cancer genes, proto-oncogenes and development, *Exp. Cell Res.,* 173, 1, 1987.

115. **Glaichenhaus, N. and Vasseur, M.,** Proto-oncogenes and differentiation, *Cancer J.,* 1, 255, 1987.

116. **Adamson, E. D.,** Oncogenes in development, *Development,* 99, 449, 1987.

117. **Atkinson, M. and Sheridan, J.,** Decreased junctional permeability in cells transformed by 3 different viral oncogenes: a quantitative video analysis, *J. Cell Biol.,* 99, 401a, 1984.

118. **Azarnia, R. and Loewenstein, W. R.,** Intercellular communication and the control of growth: alteration of junctional permeability by the src gene: a study with temperature-sensitive mutant Rous sarcoma virus, *J. Membr. Biol.,* 82, 191, 1984.

119. **Chang, C. C., Trosko, J. E., Kung, H. J., Bombick, D., and Matsumura, F.,** Potential role of the src gene product in the inhibition of gap junctional communication in NIH3T3 cells, *Proc. Natl. Acad. Sci. U.S.A.,* 82, 5360, 1985.

120. **Atkinson, M. M., Anderson, S. K., and Sheridan, J. D.,** Modification of gap junctions in cells transformed by a temperature-sensitive mutant of Rous sarcoma virus, *J. Membr. Biol.,* 9, 53, 1986.

121. **El-Fouly, M. H., Warren, S. T., Trosko, J. E., and Chang, C. C.,** Inhibition of gap junction-mediated intercellular communication in cells transfected with the human H-ras oncogene, *Am. J. Hum. Genet.,* 39, A30, 1986.

122. **Azarnia, R., Reddy, S., Kmiecki, T. E., Shalloway, D., and Loewenstein, W. R.,** The cellular src gene product regulates junctional cell to cell communication, *Science,* 239, 398, 1988.

123. **Azarnia, R. and Loewenstein, W. R.,** Polyomavirus middle T antigen down regulates junctional cell to cell comunication, *Molec. Cell. Biol.,* 7, 946, 1987.

124. **Iijima, N., Yamamoto, T., Inoue, K., Soo, S., Matsuzawa, A., Kanno, Y., and Matsui, Y.,** Experimental studies on dynamic behavior of cells in subcutaneous solid tumor (MH134-C3H/He), *Jpn. J. Exp. Med.,* 39, 205, 1969.

125. **Hamada, J., Takeichi, N., and Kobayashi, H.,** Inverse correlation between the metastatic capacity of cell clones derived from a rat mammary carcinoma and their intercellular communication with normal fibroblasts, *Gann,* 78, 11175, 1987.

126. **Nicolson, G. L., Dulski, K. M., and Trosko, J. E.,** Loss of intercellular junctional communication correlates with metastatic potential in mammary adenocarcinoma cells, *Proc. Natl. Acad. Sci. U.S.A.,* 85, 473, 1988.

127. **Land, H., Parada, L. F., and Weinberg, R. A.,** Tumorigenic conversion of primary embryo fibroblasts requires at least two cooperating oncogenes, *Nature,* 304, 596, 1983.

128. **Newbold, R. F., and Overell, R. W.,** Fibroblast immortality is a prerequisite for transformation by EJ v-Ha-ras oncogene, *Nature,* 304, 648, 1983.

129. **Spandidos, D. A. and Wilkie, N. M.,** Malignant transformation of early passage rodent cells by a single mutated human oncogene, *Nature,* 310, 469, 1984.

130. **Brown, K., Quintanilla, M., Ramaden, M., Kerr, I. B., Young, S., and Balmain, A.,** V-ras genes from Harvey and BALB murine sarcoma viruses can act as initiators of two-stage mouse skin carcinogenesis, *Cell,* 46, 447, 1986.

131. **Geiser, A. G., Der, C. J., Marshall, C. J., and Stanbridge, E. J.,** Suppression of tumorigenicity with continued expression of the c-Ha-ras oncogene in EJ bladder carcinoma-human fibroblast hybrid cells, *Proc. Natl. Acad. Sci. U.S.A.,* 83, 5209, 1986.

132. **Koi, M. and Barrett, J. C.,** Loss of tumor-suppressive function during chemically induced neoplastic progression of Syrian hamster embryo cells, *Proc. Natl. Acad. Sci. U.S.A.,* 83, 5992, 1986.

133. **Decker, R. S., Donta, S. T., Larsen, W. J., and Murray, S. A.,** Gap junctions and ACTH sensitivity in y-1 adrenal tumor cells, *J. Supramolec. Struct.,* 9, 497, 1978.

134. **Gansler, T. and Kopelovich, L.,** Effects of TPA and epidermal growth factor on the proliferation of human mutant fibroblasts in vitro, *Cancer Lett.,* 13, 315, 1981.

135. **Harrison, J. and Auersperg, N.,** Epidermal growth factor enhances viral transformation of granulosa cells, *Science,* 213, 218, 1981.

136. **Stoscheck, C. M. and King, L. E.,** Role of epidermal growth factor in carcinogenesis, *Cancer Res.,* 46, 1030, 1986.

137. **Lipton, A., Kepner, N., Rogers, C., Witkoski, E., and Leitzel, K.,** A mitogenic factor for transformed cells from human platelets, in *Interactions of Platelets and Tumor Cells,* Jamieson, G. A., Ed., Alan R. Liss, New York, 1982, 233.

138. **Rose, S. P., Stahn, R., Passovoy, D. S., and Herschman, H.,** Epidermal growth factor enhancement of skin tumor induction in mice, *Experientia,* 32, 913, 1976,

139. **Takizawa, S. and Hirose, F.,** Role of testosterone in the development of radiation-induced prostate carcinoma in rats, *Gann,* 69, 723, 1978.

140. **Yager, J. D. and Yager, R.,** Oral contraceptive steroids as promoters of hepatocarcinogenesis in female Sprague-Dawley rats, *Cancer Res.,* 40, 3680, 1980.

141. **Yosida, H., Fukunishi, R., Kato, Y., and Matsumoto, K.,** Progesterone-stimulated growth of mammary carcinomas induced by 7,12-dimethylbenz(a)anthracene in neonatally androgenized rats, *J. Natl. Cancer Inst.,* 65, 823, 1980.

142. **Chester, J. F., Gaissert, H. A., Ross, J. S., and Malt, R. A.,** Pancreatic cancer in the Syrian hamster induced by N-nitrosobis (2-oxopropyl)-amine cocarcinogenic effect of epidermal growth factor, *Cancer Res.,* 46, 2954, 1986.

143. **Trosko, J. E. and Chang, C. C.,** Adaptive and nonadaptive consequences of chemical inhibition of intercellular communication, *Pharmacol. Rev.,* 36, 137, 1984.

144. **Binggeli, R. and Weinstein, R. C.,** Membrane potentials and sodium channels: hypothesis for growth regulation and cancer formation based on changes in sodium channels and gap junctions, *J. Theoret. Biol.,* 123, 377, 1986.

145. **Spray, D. C. and Bennett, M. V. L.,** Physiology and pharmacology of gap junctions, *Annu. Rev. Physiol.,* 47, 281, 1985.

146. **Takeda, A., Hashimoto, E., Yamura, H., and Shinazu, T.,** Phosphorylation of liver gap junction protein by protein kinase C, *FEBS Lett.,* 210, 169, 1987.

147. **Wiener, E. C. and Loewenstein, W. R.,** Phosphorylation of lens membrane with a cyclic AMP-dependent protein kinase purified from the bovine lens, *Nature,* 305, 433, 1983.

148. **Saez, J. C., Spray, D. C., Nairn, A. C., Hertzberg, E., Greengard, P., and Bennett, M. V. L.,** C-AMP increases junctional conductance and stimulates phosphorylation of the 27-kDa principal gap junction polypeptide, *Proc. Natl. Acad. Sci. U.S.A.,* 83, 2473, 1986.

149. **Saez, J. C., Bennett, M. V. L., and Spray, D. C.,** Carbon tetrachloride at hepatoxic levels blocks reversibly gap junctions between rat hepatocytes, *Science,* 236, 967, 1987.

150. **Warngard, L., Flodstrom, S., Ljungquist, S., and Ahlborg, U. G.,** Interaction between quercetin, TPA and DDT in the V79 metabolic cooperation assay, *Carcinogenesis,* 8, 1201, 1987.

151. **Aylsworth, C. F., Trosko, J. E., Chang, C. C., Benjamin, K., and Lockwood, E.,** Synergistic inhibition of metabolic cooperation by oleic acid or 12-o-tetradecanoylphorbol-13-acetate and dichlorodiphenyl-trichlorethane in Chinese hamster cells: implications of a role for protein kinase C in the regulation of gap junctional intercellular communication, *Cell Biol. Toxicol.,* in press.

152. **Meda, P., Bruzzone, R., Knodel, S., and Orci, L.,** Blockage of cell to cell communication with pancreatic acini is associated with increased basal release of amylase, *J. Cell Biol.,* 103, 475, 1986.

153. **Davidson, J. S., Baumgarten, I. M., and Harley, E. H.,** Reversible inhibition of intercellular junctional communication by glycyrrhetinic acid, *Biochem. Biophys. Res. Commun.,* 29, 1986.

154. **Klein, G.,** The approaching era of the tumor suppressor genes, *Science,* 238, 1539, 1987.

Chapter 8

TERATOGENS AND CELL-TO-CELL COMMUNICATION

Frank Welsch

TABLE OF CONTENTS

I. INTRODUCTION

Cell interactions in embryos derived from a large diversity of phyla of the animal kingdom have been studied by developmental biologists for many decades. The interest in this specific topic has its foundation in the desire to understand more about the factors that direct the development of a single cell of the fertilized ovum to a precisely structured multicellular organism. At the beginning of this century it became apparent that cell interactions during embryogenesis were critical for normal differentiation. Morphogenetic communication was first systematically explored by Spemann in his classic transplantation experiments in which he studied interactions between the optic vesicle and the lens.[1] Later the phenomenon became also known as "morphogenetic tissue interactions" or "inductive tissue interactions".

It is generally accepted that normal, genetically controlled morphogenesis will only occur if cells can communicate with one another in a developmentally appropriate manner.[2] Because of the complexity of embryonal development, progress has been slow. However, the availability of new investigational methods in molecular biology during the last decade has laid the foundation for progress in elucidating events of normal embryogenesis. This knowledge is needed to examine abnormal differentiation as induced by teratogens. Mechanisms of chemically induced dysmorphogenesis are being studied in a variety of model systems where, relatively speaking, there exists the best data base about normal development. However, there is no common approach, nor is one to be expected in view of the multitude of mechanisms that has been speculated to induce birth defects.[3] Among a dozen or so intracellular and extracellular target sites of chemical insult (Figure 1) thought to be involved in the chain of events leading to birth defects are alterations of membrane function that might cause failure of critical cell-cell interactions.[4]

There are numerous ways by which cells may interact with one another during morphogenesis (Figure 2), and several have been studied intensely by developmental biologists. The accumulated data have established that many morphogenetic processes, including cell migration and recognition, tissue induction, cell morphology, response to hormonal stimuli and formation as well as maintenance of cell-to-cell channels, depend on the integrity of the cell membrane. While there is a considerable data base which describes the importance of the latter, little information relevant to chemical teratogen-induced alterations of membrane function exists. Many of the model systems to study morphogenetic interactions during embryogenesis were recently reviewed with emphasis on perturbation by chemical teratogens wherever such experiments had been done.[5] This chapter will focus on only one of the modes of cell-cell interactions depicted schematically in Figure 2, and that is the one which involves specialized area of membrane-membrane contact where junctional complexes of the gap junction-type are formed.

II. GAP JUNCTIONS

A. HISTORICAL SKETCH

Low resistance intercellular junctions between electrically nonexcitable cells were first observed with electrophysiological methods in 1964.[6] Three years later a cell membrane organelle was recognized by its ultrastructure and appeared to connect the

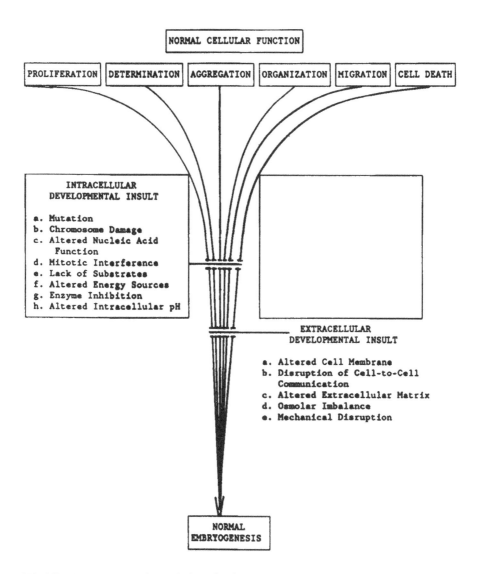

FIGURE 1. Potential sites of perturbation of embryogenesis by chemical teratogens. This schematic drawing gives an overview of the many cellular processes and interactions that need to occur with precise timing to result in normal embryogenesis. Chemicals that are capable of interfering with developmental events are distinguished by intra- and extracellular target sites and their efforts on cellular homeostasis. (From Welsch, F., in *Issues and Reviews in Teratology*, Vol. 5, Kalter, H., Ed., Plenum Press, New York, in press. With permission.)

intracellular compartments of cells in close apposition. This specialized area had building block contributions (called hemichannels or connexons) from both cell membranes of adjacent cells, and its ultrastructural appearance seemed compatible with the hydrophilic cell-to-cell channel suggested by cell coupling observations. Revel and Karnovsky[7] introduced the term "gap junction" to describe aggregates of cell-cell channels coincident with the characteristic 2- to 4-nm gap between apposed

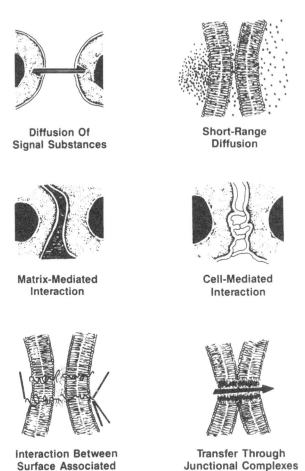

Diffusion Of
Signal Substances

Short-Range
Diffusion

Matrix-Mediated
Interaction

Cell-Mediated
Interaction

Interaction Between
Surface Associated
Components

Transfer Through
Junctional Complexes

FIGURE 2. Cell-cell interactions in embryogenesis. Signal substances are transmitted by different modes through morphogenetic cell interactions between cells at some distance from one another and in close proximity to one another. Only the effects of chemicals on transfer through junctional complexes of the gap junction type is the subject of this chapter. (From Welsch, F., in *Biochemical Mechanisms and Regulation of Intercellular Communication*, Milman, H. A. and Elmore, E., Eds., Princeton Scientific Publishers, Princeton, 1987, 116. With permission.)

cell membranes. Remarkable progress has since been made in defining the structure, distribution, permeability characteristics, and biochemical makeup of gap junctions.[8-14] Several specific treatments have been found that cause either opening or closing, i.e., gating, of cell-to-cell channels. Some of such treatments are rapid in onset, while others operate over a longer time course to affect the formation or disappearance of the channels.[18,20] Elucidation of the actual biological functions of gap junctions proved to be difficult, although the term "junctional (or gap junctional) communication" has been accepted in the literature for some time. However, gap junction proteins from several species have recently been purified and the corresponding cDNA sequences cloned[14] (see Chapter 1). As shall be discussed, these advances in the biochemical and molecular composition of the gap junction have laid the foundation for significant gains in our understanding of the roles of gap junction in embryogenesis.

B. PHYSIOLOGICAL FUNCTIONS OF GAP JUNCTIONS IN EMBRYOGENESIS

The accumulated evidence that links gap junctions with events in embryo development was recently summarized by Guthrie.[15] The timing of appearance, presence, and disappearance of gap junctions precedes or coincides with specific developmental events, which lends support to the idea that communication via cell-to-cell channels provides a specific pathway for intercellular signals of a transient nature. Regulation of cell growth, morphogenetic gradient and pattern formation, and differentiation through the intercellular transmission of as yet unidentified regulatory molecular signals have been invoked as some of the functions of gap junctions.[6,16-19]

As regards embryogenesis, cell-to-cell channels have commonly been studied in embryos of lower vertebrates because of the large cell sizes that allow simultaneous multiple micropipet impalements required for ionic coupling measurement.[16,20] In mammals gap junctional communication has been measured by ionic coupling and with fluorescent dye coupling methods in pre- and postimplantation mouse embryos.[21,22] One speculation about the function of gap junctions is that they may facilitate the buildup of chemical gradients during development.[16,17] The development of regional patterns was observed, which led to the formation of "communication compartments". Cells in a given compartment are well coupled with one another, whereas dye coupling is severely restricted between various compartments.[21] Recent investigations have further refined coupling boundaries in egg cylinder stage mouse embryos. This is the developmental stage at which the overall body plan of the mammalian embryo is laid down and critical determinations about separate germ layers are made. Differences were observed between the extent of fluorescent dye as opposed to ionic coupling.[22] Dye coupling occurred within all three germ layers, but there were severe restrictions to transfer between germ layers. Within a given germ layer dye spread was restricted to specific regions of the embryo. These observations suggest a subdivision within each germ layer of cells into smaller communication compartments. The most striking finding was that of "box-like" dye coupling compartments within a single germ layer with slight overlap into adjacent layers. Kalimi and Lo[22] also used ionic coupling measurements to pursue the nature of dye coupling observed in the box-like compartments. It was then observed that at least a low level of ionic coupling between germ layers was present even in the absence of detectable dye transfer. Overall it was concluded[22] that the mouse embryo findings together with those from previous work in insects[23] allow the speculation that communication compartments coincide with "developmental compartments".

C. PERTURBATION OF GAP JUNCTIONS
1. General Considerations

The modulation of cell-cell channel permeability by altering conditions such as intracellular pH, calcium, or cyclic nucleotide concentrations was established many years ago.[18] The impact of these factors and the effect of various pharmacological agents is the subject of several recent reviews.[20,24,25] In the past few years the perturbation of gap junctional intercellular communication by chemicals that are commonly considered toxicants and/or environmental contaminants has received increasing attention[26-28] (see Chapter 6).

2. Interference with Gap Junctions in Embryogenesis

For a long time a major shortcoming affecting experimental design in efforts to explore the functional role of gap junctions, including that in prenatal development, was the lack of specific and reversible probes to interfere with junctional permeability. While the experimental evidence that gap junctions were important in embryonal development was already quite convincing in the late 1970s, more definitive proof remained elusive. In reviewing the data, Lo[21] stated that "in order to discern the role of cell-cell communication in embryogenesis, it would be necessary to be able to specifically inhibit gap junctional communication and study the subsequent effects on development."

Major progress came when, after many years of effort in several laboratories, the protein that makes up gap junctions in rat liver was prepared in purity and quantity sufficient to produce antibodies.[29] Affinity-purified polyclonal antibodies exhibited dramatic inhibitory effects on gap junctional conductance in isolated pairs of hepatocytes.[8,30] A turning point for embryogenesis was the injection of similar antibodies into a specific cell of *Xenopus* embryos at the 8-cell stage of development.[31] Ionic coupling and fluorescent dye coupling with neighboring cells remained impaired for an extended time. Injections of the antibodies at the 32-cell stage followed immediately by Lucifer Yellow showed that dye coupling via existing cell-cell channels was disrupted within minutes. In many instances malformed embryos resulted with grossly visible structural deviation in a body region that derived its cells from the cell injected with the antibodies.[31,32] This outcome has high relevance to the postulated role of gap junctional communication in embryogenesis. Antibody probes to gap junctions have also proven very valuable in studies of other tissues and affect gap junctional conductance.[8,30] Antibodies against rat or mouse liver gap junction protein were used successfully to demonstrate, by indirect immunofluorescence, the presence of gap junctions in tissues of homologous and heterologous species.[33,34] The striking effects in *Xenopus* embryos revealed that antibodies against rat liver gap junction protein interfered with gap junction function. This and other observations indicate a considerable degree of conservation in the protein across species and phyla.

More support for the role of gap junctions in development and pattern formation derives from the elegant grafting studies in Hydra conducted by Fraser and associates.[35] Antibodies from the same source as in the *Xenopus* experiments were used to localize gap junctions in *Hydra attenuata* by immunocytochemistry. Furthermore the effects of the antibodies on intercellular communication were measured by ionic and dye coupling, both of which were inhibited. Specific developmental phenomena associated with regeneration and patterning were also disrupted. The data obtained are compatible with the notion that transmission of a diffusible low molecular weight substance via gap junctions plays a role in the inhibition of regeneration of a second head/axis. A segment of Hydra capable of complete regeneration was grafted to the midgastric region of a host. Normally a diffusible head inhibition factor from the existing head region prevents the regenerative formation of a second head/axis in the grafted segment. However, this process was significantly inhibited when the coelenterate's cells had been loaded with the antibodies prior to grafting, and more of the grafted specimens developed a second head/axis. The investigators attributed this effect to disruption of diffusion of the putative head formation inhibition factor which reaches the grafted piece from the existing head region via gap junctions.[35]

3. Chemical Teratogens and Gap Junctions
a. Origin of the Hypothesis

In vitro studies on the mechanism of action of tumor-promoting chemicals revealed their ability to inhibit metabolic cooperation via gap junctions. This phenomenon and the probability of some commonalities between disrupted cellular differentiation in carcinogenesis and teratogenesis led Trosko and his associates to postulate that disruption of cell-cell communication might be a mechanism also applicable to chemical teratogens.[36] Since then the subject has undergone some experimental evaluation with compounds that are generally recognized as causing birth defects.[37-39] The concept and the circumstantial evidence supporting it have been comprehensively reviewed.[40]

b. Effects of Teratogens

We have previously examined a variety of chemical teratogens for their effects on intercellular communication with three methods of very different instrumentation complexity.[37-39] In cultured cells the assay endpoints were indicative of intercellular communication via gap junctions. The V79 Chinese hamster cell metabolic cooperation assay revealed that several teratogens interfered with this phenomenon.[37] Thereafter quantitative autoradiography in a different cell line was adopted to enhance the sensitivity of the method used to assess gap junctional communication.[38] This allowed shortening of the teratogen exposure time from 3 d in the V79 metabolic cooperation assay to 3 h, a period that seemed more compatible with the time frame in which critical developmental events occur *in vivo* which can be disrupted by chemical teratogens. However, the experimental data are inadequate to draw any conclusions about the specific site of action of chemicals on gap junctions. While the cell membrane has been postulated to be the potential target for teratogens and direct interruption of junctional communication, a vast number of biochemical events would need to be examined to pinpoint the primary site of chemically induced perturbation. It was important for our hypothesis about teratogen effects on gap junctional communication to evaluate the junctions more directly in their response to a xenobiotic. However, there is still no published method to assess the quantity of gap junctions or their characteristic protein aside from quantitative morphometry, which can be applied to determine the presence, size, and fractional area of gap junctions in relationship to the total cell surface area. The approach is labor intensive, but analysis of freeze-fracture replicas is a proven technique for quantitation of the gap junction response to physiological factors[41-43] and to a tumor-promoting agent.[44] We performed morphometric measurements under control and teratogen treatment conditions at the exact time of drug exposure at which inhibition of intercellular communication was determined. The data revealed that the model teratogen had significantly reduced the incidence of gap junctions and their fractional area.[38]

c. Studies with Differentiating Tissues

It has long been recognized that for *in vitro* studies concerning mechanisms of chemically induced abnormal development it would be desirable to use model systems that undergo some aspect of progressive development and differentiation.[45] Differentiating mammalian limbs or avian wing buds are one of the most intensely investigated tissues in developmental biology. Therefore the normal morphogenesis of the limb/wing bud is relatively well characterized.[46-49] In high-density cultures mesenchymal cells from these appendages differentiate to chondroblasts and further to chondrocytes. The latter produce cell specific collagen type II and proteoglycans that are deposited as extracellular cartilaginous matrix.[50-53] These developmental phenomena resemble those occurring *in vivo*.

There is ample evidence that during limb morphogenesis both mesenchymal cells and those differentiating from them as well as cells of the overlying epithelial cover are at certain times coupled by gap junctions. At the distal end (tip) of the limb bud the epithelial cells form a special thickened region that is known as the apical ectodermal ridge (AER). Cells of the AER have more and larger gap junctions than the mesenchymal cells[54] and are dye coupled in the intact preparation (Fallon, personal communication). Within hours after seeding single cell suspensions from limb buds, gap junctions are established between the cells, the overwhelming majority of which are made up of mesenchyme.[55,56] It has been postulated that appearance of these junctions precedes morphogenetic differentiation and is prerequisite for its occurrence. Presumably gap junctions provide conduits for the intercellular communication of a putative morphogenetic signaling substance.[56] Several investigators have invoked cyclic nucleotides in limb cell differentiation and have found that cAMP profoundly stimulates cartilage matrix synthesis[57-59] and alters cell coupling in many different cells.[60]

Therefore, primary cultures of mouse embryo limb bud mesenchyme cells seemed to fulfill several prerequisites for studies of gap junctions during cell differentiation and were selected for the next series of experiments in which dye coupling and quantitative morphometry of the junctions were conducted after identical drug exposures. Thereafter we used whole embryos whose limbs were dissected after teratogen exposure in short term culture and examined the AER. In the most recent experiments limbs from mouse embryos exposed to the model teratogen *in utero* were analyzed for incidence and fractional area of the surface in AER cells that was occupied by gap junctions.

All-*trans* retinoic acid (ATRA; generically also known as tretinoin) was used as a representative chemical teratogen, and limited comparative studies were conducted with its 13-*cis* analog, 13-*cis* retinoic acid (RA; also known as isotretinoin and marketed as drug under the trade name Accutane). These chemicals are synthetic derivatives of naturally occurring vitamin A (retinol; Figure 3). The choice of retinoids was made because hypervitaminosis A has long been known to cause perturbation of mammalian embryogenesis and induce birth defects at multiple target sites, depending on the specific dosing regimen used.[61] In pregnant laboratory animals a single dose of ATRA induces limb reduction deformities.[62] It has been known for many years that ATRA, RA, and many other retinoids inhibit differentiation of limb mesenchyme cells to chondrocytes.[51,63,64] These drug actions made ATRA suitable for the present study of its effects of gap junctional communication.

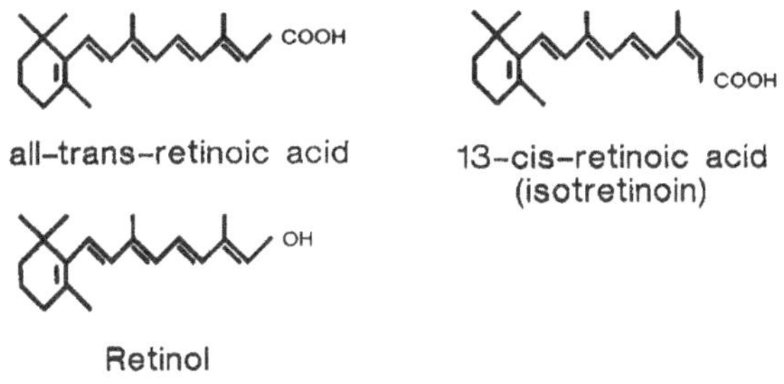

all–trans–retinoic acid

13–cis–retinoic acid
(isotretinoin)

Retinol

FIGURE 3. Structural formulas of retinol (vitamin A) and the synthetic retinoids ATRA (tretinoin) and 13-*cis*-retinoic acid (isotretinoin). These retinoids were examined for their effects on fluorescent dye coupling via gap junctions and on chondrogenic differentiation of mesenchyme cells from mouse embryo limb buds in high density cultures.

III. MATERIALS AND METHODS

A. LIMB BUD CELLS

Crl:CD-1 ICR BR (CD-1) mice of both sexes were obtained from Charles River Breeding Laboratories, Inc. (Raleigh, NC). The animal husbandry and breeding procedures were those used in earlier studies and published elsewhere in detail.[38] Nulliparous females (25 to 35 g body weight) were paired in the home cage of the male for the last 1.5 h of the dark cycle. Females with copulation plugs were presumed to be pregnant. These animals were weighed, and the next 24 h period was designated as gestation day (gd) 0.

On gd 11 (about 40 to 44 somites) dams were killed by CO_2 asphyxiation and forelimb bud cells prepared for culture as detailed before.[39] Complete culture medium consisting of CMRL-1066 (Gibco, Grand Island, NY) containing 10% fetal bovine serum, 2 mM L-glutamine and gentamicin was added to result in a final cell density of 1×10^7 cells/ml. For routine cultures 20 µl aliquots (~200,000 cells) were transferred and placed as a drop into 35-mm culture dishes. After a 2-h incubation to allow cell attachment, the resultant spot (high density or "micromass") cultures were flooded with 2 ml of complete medium. The dishes were incubated at 37°C in a humidified atmosphere of 95% air and 5% CO_2. These conditions at high cell density are conducive to cartilaginous differentiation of limb mesenchymal cells and to the production of chondrogenic matrix.[50,52,53,55] Within hours after cell explantation, gap junctions were observed.[55,56] The regions where mesenchyme cells were undergoing differentiation to chondroblasts and chondrocytes coincident with cartilage nodule formation were clearly distinguishable from other cells on the third day (i.e., about 48 h in culture). At that time the synthetic retinoids (Figure 3) were added in concentrations ranging from 0.01 µg to 10.0 µg/ml culture medium. ATRA (Sigma Chemicals, St. Louis, MO) and RA (isotretinoin; Ro 4-3780, received through the courtesy of Dr. Peter Sorter from Hoffmann-LaRoche, Nutley, NJ) were dissolved in ethanol (volume not exceeding 0.5%) and added to the culture medium.

In dishes destined to undergo chondrogenic differentiation and quantitation of cartilaginous matrix, the medium was changed 24 h later (i.e., 3 d after initiation of the cultures), and fresh medium containing the retinoids was added for 3 more days. On day 6 the cultures were stained for 2 h with Alcian Blue (0.5% Alcian Blue in 3% glacial acetic acid, pH adjusted to 1.0 with HCl). Unbound dye was removed by washing repeatedly (3 times) with acetic acid (3%; pH 1.0). The dye bound firmly to the extracellular matrix was then extracted overnight with 1.0 ml of 4 *M* guanidine HCl. An aliquot was transferred to a 96-well microtiter plate. The absorbance of Alcian Blue was determined at 600 nm with a Titertek multiskan spectrophotometer.

Dye coupling was assessed following pneumatic microinjections of picoliter volumes of Lucifer Yellow CH (Sigma Chemical) delivered from a Picospritzer 2 (General Valve Corporation, Fairfield, NJ). Dye was injected into single cells of chondrogenic foci, usually in their periphery. This is where differentiating chondroblasts are located[51] which are the population of cells coupled by gap junctions.[65,68] Dye spread into cells in direct and indirect contact with an injected donor cell was recorded by counting the number of cells with visible amounts of Lucifer Yellow 5 min later since maximal dye distribution had occurred by then and fluorescence remained stable for quite some time. In drug-treated cultures dye was injected within minutes to 1 h after exposure to retinoids to determine the effects of time and concentration on dye coupling.

Cells grown for ultrastructural examination of gap junctions were plated in 150 cm^2 culture flasks in order to obtain a sufficient number. The cell density per milliliter of medium corresponded to the one used in dye coupling studies. Following exposure to ethanol vehicle or to 0.1 and 1.0 μg ATRA per milliliter, concentrations which affected dye coupling (see below), cells were detached from the substratum with a rubber policeman and sedimented by gentle centrifugation. Cells were fixed by resuspending the pellet in 2% glutaraldehyde/2% paraformaldehyde in 0.1 *M* phosphate buffer, pH 7.2. Further preparation for freeze-fracture and preparation of replicas followed the procedure described elsewhere.[38] Replicas were surveyed in a Philipps EM 400 electron microscope at 5000- to 6500-fold magnification for membrane areas with the structural characteristics of protoplasmic membrane (P-face). These areas were photographed on 35-mm Eastman Kodak 5302 release positive film. The processed film was placed onto a light box and scrutinized with a magnifying glass (10×) to mark those frames that contained gap junctions. The film strip was then placed into a Bessler 23 C II photographic magnifier and enlarged to about 50,000 × overall. The area of all P-faces was projected onto a digitizing tablet and traced with a pen for data acquisition via an IBM PC using Sigma Scan software (Jandel Scientific, Corte Madera, CA). Once the entire area of the randomly selected P-faces had been recorded, the number of and the cell surface area occupied by gap junctions was determined.

B. WHOLE EMBRYOS *IN VITRO*

Intact embryos to be used for *in vitro* exposures to 1.0 μg ATRA/ml were obtained from dams on gd 11 (40 to 44 somite stage). The embryo culture was a modification of that employed by Goulding and Pratt[66] for embryos which are at a later developmental stage than those typically employed in rodent whole embryo culture. Conceptuses were removed from the uterus, collected in Ham's medium and dissected free of decidual swelling. Reichert's membrane was removed and a small slit was made into an avascular area of the visceral yolk sac near the head region of the embryo, which was then gently exteriorized through a hole torn into the amniotic sac. Each intact conceptus was transferred to a 30-ml bottle that contained 4 ml of Waymouth's medium with 50% fetal bovine serum and 4 mg Pentex bovine albumin per ml (Miles Scientific, Naperville, IL). The embryos were preincubated for 30 min after the dissection perturbation, and ATRA dissolved in ethanol as described for the cell cultures was then added. Before preincubation the culture bottles were gassed with 95% O_2/5% CO_2 and continuously rotated in an incubator for 1 h. The embryos were transferred into icecold phosphate buffered saline (PBS) and washed repeatedly with PBS. The forelimb buds were dissected, pooled in PBS, and fixed as in the micromass cultures.

Freeze-fracture analysis of gap junctions was specifically aimed at the epithelial cells of the AER. Therefore excess limb mesenchymal tissue was trimmed off the specimens prior to fracturing to provide access to the desired cell population.

C. EMBRYOS EXPOSED *IN VIVO*

On gd 11 dams received a single oral administration of 100 mg ATRA per kilogram, which induced limb malformations in almost 90% of the embryos.[62] In the conceptus peak concentrations of the parent compound are reached after 2 h, a time at which the levels of another embryotoxic ATRA metabolite are still increasing.[67] Mice were killed 1 and 2 h after being given the drug. Embryos were removed from the uterine horns, dropped into Hank's buffered salt solution for dissection of the forelimb buds, and prepared for freeze-fracture of the AER as described in the preceding section.

IV. RESULTS AND DISCUSSION

A. CELL CULTURE OBSERVATIONS

Several studies have revealed that cAMP enhances chondrogenic differentiation of cultured limb mesenchyme cells.[52,53] In other cells or cell lines cAMP stimulates gap junctional intercellular communication.[18,60] We determined dye coupling under control conditions and after exposure to dibutyryl cAMP (dbcAMP). Limb mesenchymal cells undergoing differentiation to chondrocytes were extensively dye coupled 48 h after explantation into high-density cell cultures. The premise that cell coupling, chondrogenic differentiation, and matrix formation are related with one another seems supported by the effects of dbcAMP. Without detailed analysis of the time course it was observed that dye coupling was significantly increased 5 h after drug addition. The number of cells coupled after 24 h in the presence of dbcAMP (i.e., between 48 and 72 h after cell explantation) was about doubled compared to controls (Table 1), and proteoglycan synthesis was more than twice as high when Alcian Blue absorption was measured 4 d after adding dbcAMP. The stimulatory effect agrees with that which was first observed by Loewenstein, who demonstrated that cAMP enhanced cell coupling and the number of cell-to-cell channels.[18,60] Our results also concur with those showing that during chondrogenic differentiation of mouse limb bud cells *in vitro*, cAMP, its analogs, or other chemical agents that elevate intracellular cAMP levels stimulate the synthesis of cartilaginous matrix.[52,57-59] Dye coupling observations in control and retinoid exposed cultures were correlated with the synthesis of extracellular cartilaginous matrix. However, an exact time match was not practical, because proteoglycan staining is not sensitive enough. Therefore Alcian Blue was quantitated 6 d after cell explantation. Adverse drug effects were apparent from the gross morphology of the cells which lacked the typical phenotype of chondrogenic differentiation. This was

TABLE 1
Effects of Dibutyl-3′,5′-Cyclic AMP (dbcAMP) on Dye
Coupling and on Chondrogenic Differentiation

Treatment	Dye coupling	Alcian Blue absorption
Control	16.00 ± 7.95	0.124 (=100%)
dbcAMP (1 mM)	31.93 ± 8.14	0.301 (241%)

Note: Limb bud mesenchyme cells were explanted in high density micromass culture on gestation day 11, and 48 h later dbcAMP was added. Dye coupling was assessed 24 h later, and the data show the number of fluorescent recipient cells ± SD resulting from injections into at least 20 cells. Alcian Blue was quantitated 6 d after initiating culture.

From Welsch, F., Stedman, D. B., and Carson, J. L., in *Approaches to Elucidate Mechanisms in Teratogenesis*, Welsch, F., Ed., Hemisphere, Washington, D.C., 1987, 247. With permission.

even more striking and obviously related to ATRA concentration once the cartilage nodules had been stained with Alcian Blue (Figure 4). As expected, both ATRA and RA inhibited chondrogenic differentiation of mesenchyme cells in a concentration-dependent manner that is reflected in the amount of dye extracted from Alcian Blue-stained proteoglycans (Table 2) and inhibited dye coupling.[39]

Lucifer Yellow injected into single (donor) cells of control cultures 48 h after explantation spread rapidly into neighboring (recipient) cells in direct (primary recipients) and indirect (secondary recipients) contact with the donor cell. Maximal fluorescent cell counts were reached within a few minutes (Figure 5, panel A) and were recorded 5 min after injection. Depending on the concentration of ATRA, dye coupling was completely blocked within 2 min by 10.0 µg/ml, while at 1.0 µg/ml, a level that is much more compatible with the chondrogenesis-inhibiting efficacy of the drug, reduction occurred more gradually with increasing exposure time (Table 3; Figure 5, Panel C). Under the light microscope no morphological effects were apparent within 1 h after ATRA addition as regards cell shape. Membrane apposition and random plasma membrane particle distribution in freeze-fracture replicas revealed no discernible differences between control and ATRA-treated cells. Quantitative morphometry of gap junctions in control cells and in those exposed to 0.1 µg ATRA per milliliter or 1.0 µg/ml was performed after drug treatment for 1 h coincident with the final dye coupling assessment. Cells from three independent replicas were fixed and prepared for freeze-fracture in a double-bind design. The codes were only revealed once all data had been assembled (Table 4). In a comparable number of fields and total P-face membrane area, the number of gap junctions was reduced in cells exposed to 0.1 µg ATRA per milliliter, but remained unchanged by 1.0 µg/ml compared to controls. However, both concentrations of ATRA significantly reduced the area of individual gap junctions, and thus the total gap junctional area, and also the percentage of the cell surface occupied by gap junctions (Table 4). These effects of ATRA resemble those observed when high-density limb cell cultures explanted on gd 12 were grown for 3 d in the presence of ATRA at the same concentrations as used here. Although there was

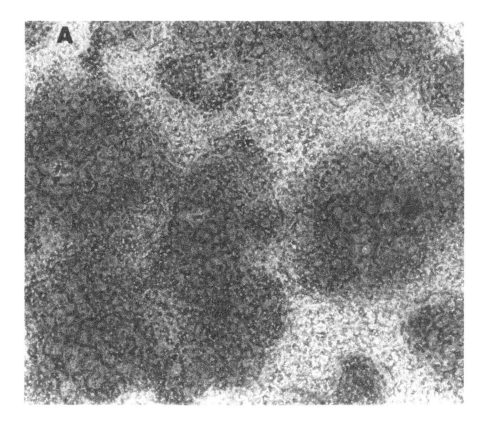

FIGURE 4. Chondrogenic differentiation of limb bud cells. Mesenchymal cells from mouse embryo limb buds were explanted into high density micromass culture on gd 11. ATRA was added 48 h later and remained present for 4 d. The photomicrographs illustrate the extent of the chondrogenic foci with their extracellular cartilaginous matrix stained darkly by Alcian Blue. Panel A is a control culture, while the others illustrate the inhibition of proteoglycan synthesis by increasing concentrations of ATRA; 0.01 (B), 0.1 (C), 1.0 (D), and 10.0 µg/ml (E). More and more of the culture dishes is occupied by cells that do not undergo chondrogenic differentiation. (Magnification × 50.) (With permission from the Chemical Industry Institute of Technology, Research Triangle Park, NC.)

an increase in cell-cell contacts, there were no gap junctions.[64] However, these observations contrasted markedly from those in limb buds removed from embryos on gd 11 and maintained intact in organ culture. Gd 11 is the developmental stage at which single-cell suspensions were prepared for the present studies. In limbs cultured intact for 3 d with ATRA present throughout, far more cell contacts of the gap junction type were present than in controls.[64] Obviously the duration of ATRA treatment under both experimental conditions is very different from our design with much shorter ATRA exposures. Nevertheless in both instances, in which high density cell cultures were exposed to ATRA, the gap junction incidence was reduced.

B. EMBRYO EXPOSURE *IN VITRO*

Quantitative gap junction morphometry in AER cells of the intact limb revealed that the junctional complement responded to ATRA very similarly to the effect obtained in high-density cultures of differentiating mesenchyme cells. It did not matter whether

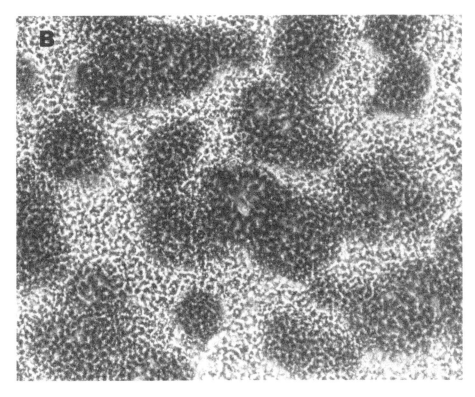

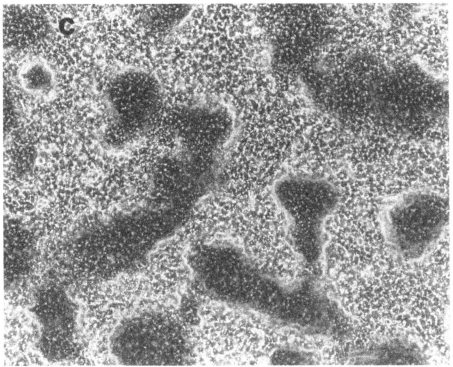

FIGURE 4 (continued)

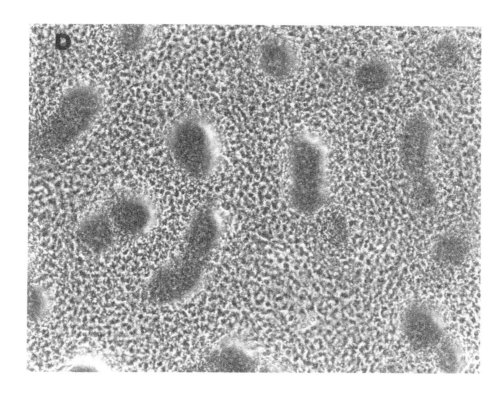

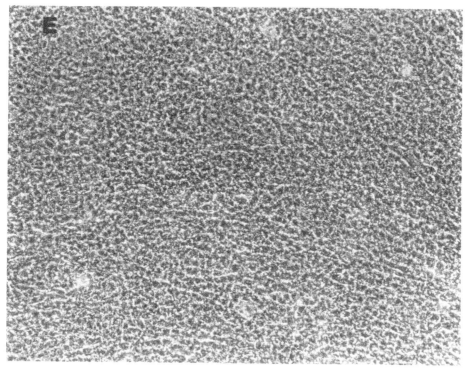

FIGURE 4 (continued)

TABLE 2
Inhibition of Chondrogenesis by Retinoids

Retinoids	0.01 μg/ml	0.1 μg/ml	1.0 μg/ml	10 μg/ml
All-*trans*-retinoic acid (ATRA)	67	18	10	4
13-Cis-RA	57	45	24	8

Note: Limb bud mesenchyme cells were explanted in high density micromass culture on gestation day 11. About 48 h later cultures were exposaed to the retinoids, and Alcian Blue staining of proteoglycans was performed 6 d after initiating culture. Values are percent of control (Alcian Blue stain).

From Welsch, F., in *Progress in Cancer Research and Therapy*, Vol. 34, Langenbach, R., Elmore, E., and Barrett, J. C., Eds., Raven Press, New York, 1988, 120. With permission.

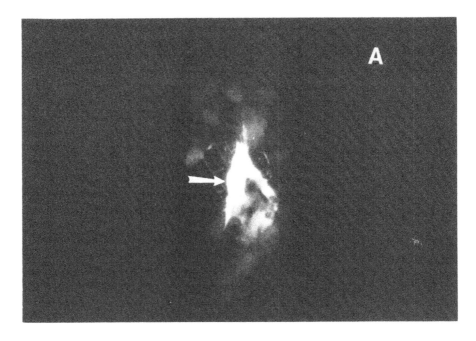

FIGURE 5. Intercellular communication by dye coupling between differentiating mesenchymal/chondrogenic cells. Mouse embryo limb bud cells were prepared for high-density micromass culture on gd 11. ATRA was added 48 h later, and microinjections of Lucifer Yellow CH were made within 1 h after drug addition. Dye spread from a single donor cell to primary and secondary recipients was quantified by counting the number of fluorescent cells 5 min later. Arrows indicate injected donor cell. (A) Control culture, fluorescent illumination; (B) brightfield of the same area on the culture dish; (C) treatment with 1.0 μg ATRA per milliliter (fluorescent illumination reveals significant inhibition of dye coupling); and (D) brightfield. (Magnification × 50.) (From Welsch, F., in *Biochemical Mechanisms and Regulation of Intercellular Communication*, Milman, H. A. and Elmore, E., Eds., Princeton Scientific Publishers, Princeton, 1987, 126. With permission.)

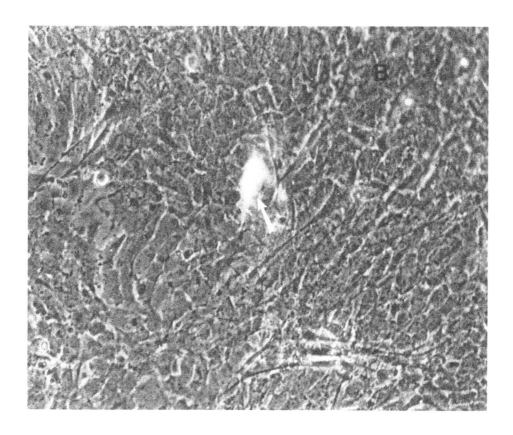

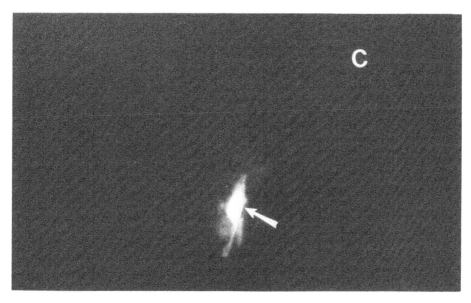

FIGURE 5 (continued)

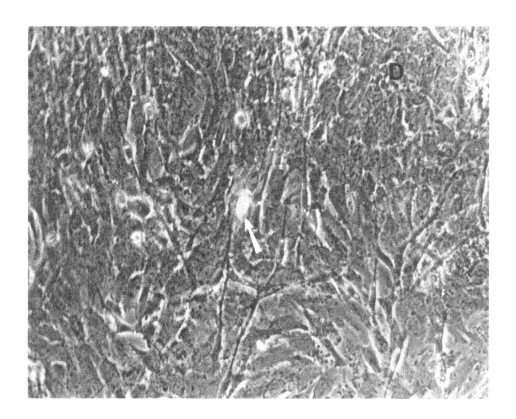

FIGURE 5 (continued)

TABLE 3
Onset of Inhibitory Effects of ATRA on Dye Coupling

	Drug exposure time					
	2 min	5 min	10 min	30 min	1 h	2 h
1 µg/ml	ND	11.46	7.00	2.00	0	0
(3.3 µ*M*)		±3.96	±3.10	±2.76		
10 µg/ml	0	0	0	ND	ND	ND
(33 µM)						

Note: Control was 14.33 ± 3.73 in this experimental series; ND, not determined. Culture
conditions were as in Table 2. About 48 h later cells were exposed to ATRA.
Numbers shown indicate the number of recipient cells labeled at various times
after microinjections of Lucifer Yellow into a single donor cell. Values are the
mean ± SE of at leat 20 injected cells.

From Welsch, F., in *Progress in Cancer Research and Therapy*, Vol. 34, Langenbach, R.,
Elmore, E., and Barrett, J. C., Eds., Raven Press, New York, 1988, 123. With permission.

TABLE 4
Incidence of Gap Junctions (Gj) in Differentiating Mouse Embryo Mesenchymal/Chondrogenic Limb Bud Cells

| | | ATRA | |
	Control	0.1 µg/ml	1.0 µg/ml
No. of fields	343	355	330
Total area (µm²)	3886	4738	4011
Total no. of gj	99	69	101
Gj/100 µm²	2.55	1.46	2.54
Area of individual gj	14.86	7.06[a]	6.51[a]
(nm², mean ± SE)	±2.37	±1.19	±0.70
Total gj area (µm²)	1.471	0.4871	0.6015
Gj area as % of total area	3.8×10^{-4}	1.00×10^{-4}	1.7×10^{-4}

Note: Culture conditions were as in Table 2. Cells were exposed to the drug for 1 h and then fixed and prepared for freeze-fracture and quantitative morphometry of gj.

[a] Significantly different ($p < 0.05$) by unpaired t-test, following $\log_{10}$ transformation of the data.

TABLE 5
Effect of ATRA on Gap Junctions (Gj) in the AER of gd 11 Limb Buds

| | *In vitro* | | *In vivo* (100 mg/kg) | | |
	Control	ATRA (1.0 µg/ml; 1 h)	Control	ATRA (1 h)	ATRA (2 h)
No. of fields	92	88	132	156	134
Total area (µm²)	1139	822	1282	1227	1496
Total no. of gj	93	29	104	65	47
Gj/100 mm²	8.17	3.52	8.11	5.30	3.14
Area of individual gj	16.0	13.0	19.0	16.0	20.0
(nm², mean ± SD)	±2.97	±2.89	±6.30	±2.90	±6.10
Total gj area (µm²)	1.48	0.377	1.97	1.04	0.94
Gj area as % of total area	1.30×10^{-3}	4.6×10^{-4}	1.54×10^{-3}	8.5×10^{-4}	6.3×10^{-4}

drug exposure of the embryos had occurred *in vitro* or *in vivo*, because in both instances there were profound effects on the gap junctions (Table 5). *In vitro* their total number was markedly lowered by 1.0 µg ATRA per milliliter present for 1 h. In turn this reduction was reflected in every parameter that expressed gap junctional area per 100 µm² P-face membrane and fractional area occupied by gap junctions vs. total P-face are (Table 5).

C. EMBRYO EXPOSURE *IN VIVO*

In AER cells of limbs from embryos exposed to the teratogenic dose of ATRA *in utero*, a significant depletion in the total number of gap junctions had occurred 1 h after drug administration and seemed more pronounced after 2 h (Table 5). The concentrations of ATRA in those limb buds are unknown, although peak levels of 1.5 µg ATRA per gram have been reported in whole embryos 2 h after dosing the dams.[67] If one were to assume even distribution in all tissues of the embryo, then this value is well within the range of the 1.0 μg ATRA per milliliter that was added to the bathing medium during *in vitro* treatment of intact embryos in our studies. As limb AER cells from the latter, those from *in vivo* exposed limbs had fewer gap junctions and concomitant reductions in total and fractional area of P-face membrane occupied by gap junctions when equivalent numbers of fields or total P-face membrane were examined (Table 5).

The number of junctions per unit plasma membrane area in AER cells was higher than in mesenchymal/chondrogenic cells as was expected based on the data reported by Fallon and Kelley.[54] The size range of junctions was also larger in limb AER than in mesenchyme derived cells. Relatively few large gap junctions (by the size standards of the population encountered in mouse limb buds) were found, together with a much larger number of small junctions, especially in ATRA-exposed specimens. Based on cell area measurements in histological sections, the total surface area of AER cells was estimated to be about 90 μm², assuming elliptoid shape. At higher magnification of P-faces and gap junctions in ATRA-exposed AER cells, there were small round areas where the P-face membrane was indented. These regions contained cell-cell channel particles as they present themselves in freeze-fracture (Figure 6, panel B). These images are compatible with those that were interpreted as indicative of endocytotic translocation into the cytoplasm when gap junctions undergo breakup and are removed from the cell surface.[10,41] In the present study the fate of the gap junctions which had disappeared from the cell surface was not further pursued. However, other investigators observed that it was on gd 11 when the cytoplasm of mouse limb mesenchymal and blastemal cells showed the highest occurrence of membrane figures designated as "coiled gap junctions". These formations were believed to result after removal of gap junctions from the cell surface membrane.[65] While AER cells were not specifically examined in that study, it appears from the observations in our control limbs as if in the present study ATRA enhanced the disappearance of gap junctions from mesenchymal/chondrogenic as well as from AER cells. The results of the quantitative morphometric measurements could not be related to the function of cell-to-cell channels since dye coupling between AER cells in intact limbs could not be measured due to the unresolved technical complexities. After 1 h ATRA treatment of high-density cultures, the total gap junctional area was reduced by more than 50%, coincident with an inhibition of about 50% by 0.1 μg/ml or abolishment of dye coupling by 1.0 μg/ml. Thus it appears that following short-term exposure to the drug there is no good correlation between morphologic presence of gap junctions and dye coupling. Cell-cell channels were no longer patent for Lucifer Yellow, yet 1 h after ATRA exposure junctions were still present, albeit in reduced number. For years intense efforts have been made with morphometric methods to assess the open or closed configuration of individual cell-cell channels. The credibility of the accumulated data is uncertain because improved tissue fixation in conjunction with cell coupling measurements has shown that there were no morphological differences between gap junctions in the low-resistance and those in the high-resistance state.[69]

Cells of different origin respond to retinoids with striking differences in the biochemical and morphological sequelae. As regards gap junctions in mouse limb buds, there were remarkable differences, depending on whether intact limbs or dissociated limb cells were exposed.[64] The outcome of ATRA treatment on gap

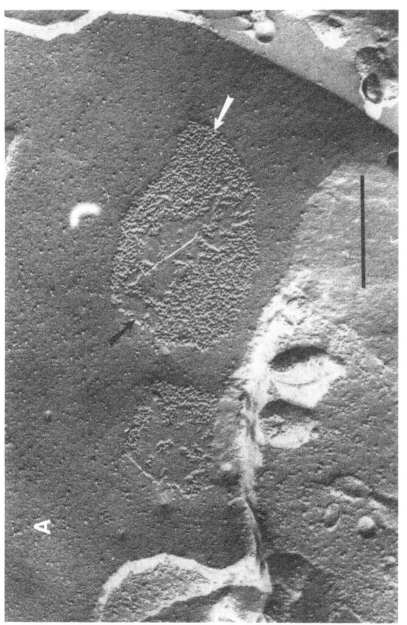

FIGURE 6. Freeze-fracture replicas of gap junctions in cells of the AER of mouse embryo limb buds. (A) Control cells show gap junctions with P-face particles (white arrow) and complementary extracellular membrane face (E-face) pits (black arrow); (B) P-face replica from ATRA- treated AER cell. Gap junction particles (arrows) can be seen and an indentation in the membrane (triangle) with gap junction particles. This formation is compatible with endocytoctic translocation of a gap junction that is breaking up. Bar size = 0.5 μm.

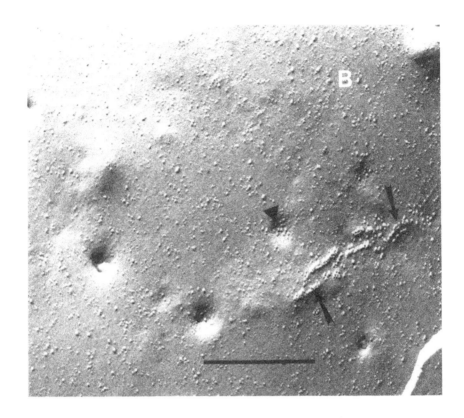

FIGURE 6 (continued)

junction function shows more consistency when various methods to assess cell coupling are compared than when morphologic endpoints are considered. The gating function and junctional conductance of cell-cell channels are rapidly reduced by ATRA,[20] and gap junctional intercellular communication is inhibited in several cell lines.[70-73] However, the biological material, treatment conditions, and drug concentrations applied are so different from the *in vitro* observations with limb mesenchymal cells that the only common outcome is inhibition of intercellular communication. Details regarding speed of onset and reversibility of ATRA's inhibitory effects in cultured limb cells and in an epithelial cell line were considered elsewhere, a comparison which emphasized that limb mesenchymal cells are very sensitive to uncoupling of gap junctions by ATRA. However, regardless of the differences in the drug's concentration that were applied to the dissimilar cells, rapid closure of cell-cell channels seemed to be the common denominator.[74] Given the multiple physiological functions of vitamin A and the complex pharmacology and toxicity of synthetic retinoids such as ATRA, it cannot be assumed without further critical examination that the drug affected only gap junctions. Thus on the basis of the present observations, ATRA cannot yet be accepted as fulfilling the criteria of a chemical probe that can specifically inhibit gap junctional communication in embryo tissues to allow assessment of the developmental consequences.[21]

It is even more difficult to reconcile the depletion of gap junctions in AER cells 1

TABLE 6
Sites of Regulation of Gap Junctions

Species	Process	Assayed parameter/technique
DNA	(1) Transcription	Message level and processing/Northern blots, nuclear transcript runons
RNA	(2) Translation	Message stability
		Protein level and stability/Northern blots
		Western blots, turnover studies
Protein	(3) Formation of functional gap junctions	Rate of formation/dye transfer, electron microscope
Gap junction	(4) Gating	Junctional conductance, permeability, phosphorylation state/biophysical techniques, dye transfer, immunoprecipitation
	(5) Disappearance of functional gap junctions	Protein level, stability, and localization/ immunoblots or immunoprecipitation of cell membrane fractions, ultrastructural immunocyto-chemistry, turnover studies, cytoskeletal disruptor effects, electrical and dye coupling
Degraded or recycled protein		

From Spray, D. C. and Saez, J. C., in *Biochemical Mechanisms and Regulation of Intercellular Communication*, Milman, H. A. and Elmore, E., Eds., Princeton Scientific Publishers, Princeton, 1987, 6. With permission.

h after treatment of mouse limb buds *in vivo* or *in vitro* with other morphologic studies reporting effects of ATRA on gap junctions in intact tissues because the target cells and treatment conditions are entirely different. For example, Prutkin applied ATRA percutaneously to chemically induced keratoacanthema on rabbit ear epidermis and observed an induction of gap junctions in cells where they normally do not occur or are very sparse.[75] During organ culture of chick embryo shank skin in the presence of ATRA for 3 d gap junction proliferation was markedly stimulated,[76] an outcome that seems similar to the limb bud organ culture results cited above.[65] In another *in vivo* study, human basal cell carcinomas were epicutaneously treated twice daily for one week with ATRA, and both the number and area of gap junctions were increased in the tumor tissue.[77]

V. PROMISING FUTURE RESEARCH DIRECTIONS

The gap junction protein is an integral membrane protein and is regulated by the same kind of control mechanisms that affect all such proteins. An overview of the experimental approaches that are now available to study different aspects of gap junctions was recently given by Spray and Saez[78] (Table 6). During the biosynthesis of gap junction protein, transcription, translation, and probably posttranslational processing are occurring before the protein reaches the membrane. There the protein becomes part of the membrane as a hemichannel and forms the complete channel by docking with another hemichannel that is contributed by the membrane of a cell in close apposition. The life cycle of the complete functional gap junction protein includes control of gating, then removal from the membrane, possibly by endocytosis, and eventual degradation.[78] Gating of existing channels constitutes a relatively rapid shift, conceivably by conformational change of the macromolecule. The other steps in the turnover of the gap function protein are considerably slower, but they determine how many assembled, functional channels are available for gating stimuli. Available estimates from pulse label experiments of cultured mouse embryo hepatocytes and other liver cell derived data have revealed half-life times of about 2 to 3 or 4 h, respectively.[79]

As with the uncoupling of intercellular dye spread between mesenchymal/chondrogenic limb bud cells by ATRA, other chemicals may exert their pharmacologic and toxic effects on gap junctions via multiple mechanisms at different levels. Our present data do not allow that distinction, but in cell pairs ATRA affected gating[20] under the stringent experimental conditions needed to measure channel gating.[78] Several experimental approaches (Table 6) would allow further dissection of the ATRA effect on limb cell gap junctions. The biochemical techniques include determinations of the effects of ATRA on gap junctional DNA transcription, mRNA, and quantitation of the protein level. Attempts concerning the latter in other cells are well underway.[80]

There are other pioneering studies revealing the directions in which the gap junction field is moving because of the tremendous progress in unraveling the molecular biology of the protein and the resultant probes that have become available to address very specific questions. For example, gap junction mRNA was injected into *Xenopus* oocytes and caused synthesis of the protein and subsequent cell coupling.[81] Other recent experiments have explored the effect of injections of antisense RNA on the early developmental processes of compaction and blastulation by the use of RNA generated from a cDNA that codes for a rat liver 32-kDa gap junction protein.[82] Severe disturbances were induced in compaction and viability of early mouse embryos coincident with inhibition of dye coupling. In toxicology, including prenatal toxicology, the evidence is now compelling that chemically perturbed gap junctional intercellular communication is a phenomenon worth investigating with methods of modern cell biology.[26,28,78]

ACKNOWLEDGMENT

The author is grateful to Mr. Donald B. Stedman for his enthusiasm during the conduct of all phases of the studies on dye coupling and gap junctions. The support of Mr. Todd M. Gambling from the Electron Microscopy facility in the Department of Pediatrics at the University of North Carolina at Chapel Hill during the preparation of all freeze-fracture replicas is very much appreciated.

REFERENCES

1. **Spemann, H.,** Über Korrelation in der Entwicklung des Auges, *Verh. Anat. Ges. (Jena)*, 15, 61, 1901.
2. **Saxen, L.,** Abnormal cellular and tissue interactions, in *Handbook of Teratology*, Vol. 2, Wilson, J. G. and Fraser, F. C., Eds., Plenum Press, New York, 1977, 171.
3. **Welsch, F.,** *Approaches to Elucidate Mechanisms in Teratogenesis*, Hemisphere, Washington, D.C., 1987, 1.
4. **Wilson, J. G.,** Current status of teratology — general principles and mechanisms derived from animal studies, in *Handbook of Teratology*, Vol. 1, Wilson, J. G. and Fraser, F. C., Eds., Plenum Press, New York, 1977, 47.
5. **Welsch, F.,** The importance of normal cellular communication during development, in *Biochemical Mechanisms and Regulation of Intercellular Communication*, Milman, H. A. and Elmore, E., Eds., Princeton Scientific, Princeton, 1987, 113.
6. **Sheridan, J. D. and Atkinson, M. M.,**Physiological roles of permeable junctions: some possibilities, *Annu. Rev. Physiol.*, 47, 337, 1985.
7. **Revel, J. P. and Karnovsky, M. J.,** Hexagonal array of subunits in intercellular junctions of the mouse heart and liver, *J. Cell Biol.*, 33, C7, 1967.
8. **Hertzberg, E. L.,** Antibody probes in the study of junctional communication, *Annu. Rev. Physiol.*, 47, 305, 1985.
9. **Bennett, M. V. L. and Spray, D. G.,** *Gap Junctions*, Cold Spring Harbor Laboratory, New York, 1985, 1.
10. **Larsen, W. J. and Risinger, M. A.,** The dynamic life histories of intercellular membrane junctions, in *Modern Cell Biology*, Vol. 4, Alan R. Liss, New York, 1985, 151.
11. **Gilula, N. B.,** Gap junctional contact between cells, in *The Cell in Contact*, Edelman, G. M. and Thiery, J.-P., Eds., John Wiley & Sons, New York, 1985, chap. 18.
12. **Green, C. R. and Gilula, N. B.,** Gap junctional communication between cells during development, in *Cellular and Molecular Control of Direct Cell Interactions*, Marthy, H.-J., Ed., Plenum Press, 1985, 337.
13. **De Mello, W. C.,** *Cell-to-Cell Communication*, Plenum Press, New York, 1987, 1.
14. **Hertzberg, E. L. and Johnson, R.,** *Gap junctions*, in *Modern Cell Biology*, Vol. 7, Alan R. Liss, New York, 1988, 1.
15. **Guthrie, S. C.,** Intercellular communication in embryos, in *Cell-to-Cell Communication*, De Mello, W. C., Ed., Plenum Press, New York, 1987, chap. 7.
16. **Caveney, S.,** The role of gap junctions in development, *Annu. Rev. Physiol.*, 47, 319, 1985.
17. **Wolpert, L.,** Gap junctions: channels for communication in development, in *Intercellular Junctions and Synapses*, Feldman, J., Gilula, N. B., and Pitts, J. D., Eds., Chapman & Hall, London, 1978, 83.
18. **Loewenstein, W. R.,** Junctional intercellular communication: the cell-to-cell membrane channel, *Physiol. Rev.*, 61, 829, 1981.
19. **Revel, J. P., Yancey, S. B., and Nicholson, B. J.,** Biology of gap junction molecules and development, in *The Cell in Contact*, Edelman, G. M. and Thiery, J.-P., Eds., John Wiley & Sons, New York, 1985, chap. 19.
20. **Spray, D. C. and Bennett, M. V. L.,** Physiology and pharmacology of gap junctions, *Annu. Rev. Physiol.*, 47, 281, 1985.
21. **Lo, C. W.,** Gap junctions in development, in *Development of Mammals*, Vol. 4, Johnson, M. H., Ed., Elsevier, New York, 1980, 39.
22. **Kalimi, G. H. and Lo, C. W.,** Communication compartments in the gastrulating mouse embryo, *J. Cell Biol.*, 107, 241, 1988.
23. **Lo, C. W.,** Communication compartmentation and pattern formation in development, in *Gap Junctions*, Bennett, M. V. L. and Spray, D. C., Eds., Cold Spring Harbor Laboratory, New York, 1985, 251.
24. **Spray, D. C., White, R. L., De Carvalho, A. C., Harris, A. L., and Bennett, M. V. L.,** Gating of gap junction channels, *Biophys. J.*, 45, 219, 1984.
25. **De Mello, W. C.,** Modulation of junctional permeability, in *Cell-to-Cell Communication*, De Mello, W. C., Ed., Plenum Press, New York, 1987, chap. 2.

26. Saez, J. C., Bennett, M. V. L., and Spray, D. C., Carbon tetrachloride at hepatotoxic levels blocks gap junctions between rat hepatocytes, *Science*, 236, 967, 1987.

27. Milman, H. A. and Elmore, E., *Biochemical Mechanisms and Regulation of Intercellular Communication*, Princeton Scientific Publishing, Princeton, 1987, 1.

28. Toxicological Implications of Altered Gap Junctional Intercellular Communication, abstracts and oral presentations, Michigan State University, Center for Environmental Toxicology, September 28 to 30, 1988.

29. Hertzberg, E. L., A detergent-independent procedure for the isolation of gap junctions from rat liver, *J. Biol. Chem.*, 259, 9936, 1984.

30. Hertzberg, E. L., Spray, D. C., and Bennett, M. V. L., Reduction of gap junctional conductance by microinjection of antibodies against the 27-kDa liver gap junction polypeptide, *Proc. Natl. Acad. Sci. U.S.A.*, 82, 2412, 1985.

31. Warner, A. E., Guthrie, S. C., and Gilula, N. B., Antibodies to gap junctional protein selectively disrupt junctional communication in the early amphibian embryo, *Nature*, 311, 127, 1984.

32. Warner, A. E., Antibodies to gap junction protein: probes for studying cell interactions during development, in *Gap Junctions*, Bennett, M. V. L. and Spray, D. C., Eds., Cold Spring Harbor Laboratory, New York, 1985, 275.

33. Dermietzel, R., Leibstein, A., Frixen, U., Janssen-Timmen, U., Traub, O. and Willecke, K., Gap junctions in several tissues share antigenic determinants with liver gap junctions, *EMBO J.*, 3, 2262, 1984.

34. Hertzberg, E. L. and Skibbens, R. V., A protein homologue to the 27,000 Dalton liver gap junction protein is present in a wide variety of species and tissues, *Cell*, 39, 61, 1984.

35. Fraser, S. E., Green, C. R., Bode, H. R., and Gilula, N. B., Selective disruption of gap junctional communication interferes with a patterning process in hydra, *Science*, 237, 49, 1987.

36. Trosko, J. E., Chang, C. C., and Netzloff, M., The role of inhibited cell-cell communication in teratogenesis, *Teratogen. Carcinogen. Mutagen.*, 2, 31, 1982.

37. Welsch, F. and Stedman, D. B., Inhibition of metabolic cooperation between Chinese hamster V79 cells by structurally diverse teratogens, *Teratogen. Carcinogen. Mutagen.*, 4, 285, 1984.

38. Welsch, F., Stedman, D. B., and Carson, J. L., Effects of a teratogen on ^{3}H-uridine nucleotide transfer between human embryonal cells and on gap junctions, *Exp. Cell Res.*, 159, 91, 1985.

39. Welsch, F., Stedman, D. B., and Carson, J. L., Teratogen interference with cell interactions: cell-to-cell channel disruption as a potential mechanism of abnormal development, in *Approaches to Elucidate Mechanisms in Teratogenesis*, Welsch, F., Ed., Hemisphere, Washington, D.C., 1987, chap. 14.

40. Loch-Caruso, R. and Trosko, J. E., Inhibited intercellular communication as a mechanistic link between teratogenesis and carcinogenesis, *CRC Crit. Rev. Toxicol.*, 16, 157, 1985.

41. Larsen, W. J., Relating the population dynamics of gap junctions to cellular function, in *Gap Junctions*, Bennett, M. V. L. and Spray, D. C., Eds., Cold Spring Harbor Laboratory, New York, 1985, 289.

42. Larsen, W. J., Wert, S. E., and Brunner, G. D., A dramatic loss of gap junctions is correlated with germinal vesicle breakdown in rat oocytes, *Devel. Biol.*, 113, 517, 1986.

43. Larsen, W. J., Wert, S. E., and Brunner, G. D., Differential modulation of rat follicle cell gap junction populations, *Devel. Biol.*, 122, 61, 1987.

44. Yancey, S. B., Edens, J. E., Trosko, J. E., Chang, C. C., and Revel, J. P., Decreased incidence of gap junctions between Chinese hamster V79 cells upon exposure to the tumor promoter 12-O-tetradecanoylphorbol-13-acetate, *Exp. Cell Res.*, 139, 329, 1982.

45. Wilson, J. G., Survey of *in vitro* systems: their potential use in teratogenicity screening, in *Handbook of Teratology*, Vol. 4, Wilson, J. G. and Fraser, F. C., Eds., Plenum Press, New York, 1977, chap. 5.

46. Kelley, R. O. and Fallon, J. F., A freeze-fracture and morphometric analysis of gap junctions of limb bud cells: initial studies on a possible mechanism for morphogenetic signalling during development, in *Limb Development and Regeneration, Part A*, Fallon, J. F. and Caplan, A. I., Eds., Alan R. Liss, New York, 1983, 119.

47. Kelley, R. O., Fallon, J. F., and Kelly, R. E., Jr., Vertebrate limb morphogenesis, in *Issues and Reviews in Teratology*, Vol. 2, Kalter, H., Ed., Plenum Press, New York, 1984, 219.

48. **Milaire, J. and Rooze, M.,** Hereditary and induced modifications of the normal necrotic patterns in the developing limb buds of the rat and mouse: facts and hypotheses, *Arch. Biol.* (Bruxelles), 94, 459, 1983.

49. **Milaire, J. and Mulnard, J.,** Histogenesis in 11-day mouse embryo limb buds explanted in organ culture, *J. Exp. Zool.,* 232, 359, 1984.

50. **Umansky, R.,** The effect of cell population density on the developmental fate of reaggregating mouse limb bud mesenchyme, *Devel. Biol.,* 13, 31, 1966.

51. **Arey-Lewis, C., Pratt, R. M., Pennypacker, J. P., and Hassell, J. R.,** Inhibition of limb chondrogenesis *in vitro* by vitamin A: alterations in cell surface characteristics, *Devel. Biol.,* 64, 31, 1978.

52. **Solursh, M., Reiter, R. S., Ahrens, P. B., and Vertel, B. M.,** Stage- and position-related changes in chondrogenic response of chick embryonic wing mesenchyme to treatment with dibutyryl cyclic AMP, *Devel. Biol.,* 83, 9, 1981.

53. **Solursh, M.,** Cell-cell interactions and chondrogenesis, in *Cartilage,* Vol. 2, Hall, B. K., Ed., Academic Press, New York, 1983, 121.

54. **Fallon, J. F. and Kelley, R. O.,** Ultrastructural analysis of the apical ectodermal ridge during vertebrate limb morphogenesis, *J. Embryol. Exp. Morphol.,* 41, 223, 1977.

55. **Merker, H.-J., Zimmermann, B., and Grundmann, K.,** Differentiation of isolated blastemal cells from limb buds into chondroblasts, in *Tissue Culture in Medical Research* (II), Richards, R. J. and Rajan, K. T., Eds., Pergamon Press, Oxford, 1980, 31.

56. **Merker, H.-J., Zimmermann, B., and Barrach, H.-J.,** The significance of cell contacts for the differentiation of the skeletal blastema, *Acta Biol. Hung.,* 35, 195, 1984.

57. **Ho, W. C., Greene, R. M., Shanfeld, J., and Davidovitch, Z.,** Cyclic nucleotides during chondrogenesis: concentration and distribution *in vivo* and *in vitro, J. Exp. Zool.,* 224, 321, 1982.

58. **Kosher, R. A.,** Relationship between the AER, extracellular matrix, and cyclic AMP in limb development, in *Limb Development and Regeneration, Part A,* Fallon, J. F. and Caplan, A. I., Eds., Alan R. Liss, New York, 1983, 279.

59. **Kosher, R. A. and Gay, S. W.,** The effect of prostaglandins on the cyclic AMP content of limb mesenchymal cells, *Cell Differ.,* 17, 159, 1985.

60. **Loewenstein, W. R.,** Regulation of cell-to-cell communication by phosphorylation, *Biochem. Soc. Symp.,* 50, 43, 1985.

61. **Howard, W. B. and Willhite, C. C.,** Toxicity of retinoids in humans and animals, *J. Toxicol. Toxin Rev.,* 5, 55, 1986.

62. **Kochhar, D. M., Penner, J. D., and Tellone, C. I.,** Comparative teratogenic activities of two retinoids: effects on palate and limb development, *Teratogen. Carcinogen. Mutagen.,* 4, 377, 1984.

63. **Kistler, A.,** Inhibition of chondrogenesis by retinoids: limb bud cell cultures as a test system to measure the teratogenic potential of compounds?, in *Concepts in Toxicology,* Vol. 3, Homburger, F., Ed., Karger, Basel, 1985, 86.

64. **Zimmermann, B. and Tsambaos, D.,** Retinoids inhibit the differentiation of embryonic-mouse mesenchymal cells *in vitro, Arch. Dermatol. Res.,* 277, 98, 1985.

65. **Zimmermann, B.,** Assembly and disassembly of gap junctions during mesenchymal cell condensation and early chondrogenesis in limb buds of mouse embryos, *J. Anat.,* 138, 351, 1984.

66. **Goulding, E. H. and Pratt, R. M.,** Isotretinoin teratogenicity in mouse whole embryo culture, *J. Craniofac. Gen. Develop. Biol.,* 6, 99, 1986.

67. **Creech-Kraft, J., Kochhar, D. M., Scott, W. J., and Nau, H.,** Low teratogenicity of 13-cis-retinoic acid (isotretinoin) in the mouse corresponds to low embryo concentrations during organogenesis: comparison to the all-trans isomer, *Toxicol. Appl. Pharmacol.,* 87, 474, 1987.

68. **Zimmermann, B. and Tsambaos, D.,** Evaluation of the sensitive step of inhibition of chondrogenesis by retinoids in limb mesenchymal cells *in vitro, Cell Differ.,* 17, 95, 1985.

69. **Hanna, R. B., Ornberg, R. L., and Reese, T. S.,** Structural details of rapidly frozen gap junctions, in *Gap Junctions,* Bennett, M. V. L. and Spray, D. C., Eds., Cold Spring Harbor Laboratory, New York, 1985, 23.

70. **Wälder, L. and Lützelschwab, R.,** Effects of 12-O-tetradecanoylphorbol-13-acetate (TPA), retinoic acid and diazepam on intercellular communication in a monolayer of rat liver epithelial cells, *Exp. Cell Res.,* 152, 66, 1984.

71. **Davidson, J. S., Baumgarten, I. M., and Harley, E. H.,** Effects of 12-O-tetradecanoylphorbol-13-acetate and retinoids on intercellular communication measured with a citrulline incorporation assay, *Carcinogenesis,* 6, 645, 1985.

72. **Pitts, J. D., Hamilton, A. E., Kam, E., Burk, R. R., and Murphy, J. P.,** Retinoic acid inhibits junctional communication between animal cells, *Carcinogenesis,* 7, 1003, 1986.

73. **Mehta, P. P., Bertram, J. S. and Loewenstein, W. R.,** Growth inhibition of transformed cells correlates with their junctional communication with normal cells, *Cell,* 44, 187, 1986.

74. **Welsch, F.,** Disruption of cell-cell channels as an event common to the potential mechanism of action of some chemicals with teratogenic and carcinogenic activity, in *Progress in Cancer Research and Therapy,* Vol. 34, Langenbach, R., Elmore, E., and Barrett, J. C., Eds., Raven Press, New York, 1988, 113.

75. **Prutkin, L.,** Mucous metaplasia and gap junctions in the vitamin A acid-treated skin tumor, keratoacanthoma, *Cancer Res.,* 33, 364, 1975.

76. **Elias, P. M. and Friend, D. S.,** Vitamin A-induced mucous metaplasia — an *in vitro* system for modualting tight and gap junctions differentiation, *J. Cell, Biol.,* 68, 173, 1976.

77. **Elias, P. M., Grayson, S., Caldwell, T. M., and McNutt, N. S.,** Gap junction proliferation in retinoic acid-treated human basal cell carcinoma, *Lab. Invest.,* 42, 469, 1980.

78. **Spray, D. C. and Saez, J. C.,** Agents that regulate gap junctional conductance: sites of action and specificities, in *Biochemical Mechanisms and Regulation of Intercellular Communication,* Milman, H. A. and Elmore, E., Eds., Princeton Scientific, Princeton, 1987, 1.

79. **Willecke, K., Traub, O., Look, J., Stutenkemper, R., and Dermietzel, R.,** Different protein components contribute to the structure and formation of hepatic gap functions, in *Modern Cell Biology,* Vol. 7, Hertzberg, E. L. and Johnson, R., Eds., Alan R. Liss, New York, 1988, 41.

80. **Lomneth, C. S., Wilfinger, W. W., and Larsen, W. J.,** Quantitation of a gap junction protein in a variety of tissues by radioimmunoassay, Int. Conf. Gap. Junctions (abstr.), Pacific Grove, CA, July 6—10, 1987.

81. **Dahl, G., Miller, T., Paul, D., Voellmy, R., and Werner, R.,** Expression of functional cell-cell channels from cloned rat liver gap junction complementary DNA, *Science,* 236, 1290, 1987.

82. **Bevilacqua, A., Loch-Caruso, R., and Erickson, R.,** Altered development of mouse embryos by antisense gap junction RNA, Abstracts, Symp. Toxicological Implications of Altered Gap Junctional Intercellular Communication, Michigan State University, East Lansing, MI, September 28 to 30, 1988.

INDEX

Printed and bound by CPI Group (UK) Ltd, Croydon, CR0 4YY

22/10/2024

01777633-0018